Praise for *Training Data for Machine Learning*

Anthony Sarkis offers a clear overview of the machine learning training process, including the nuances of annotation workflows. A valuable read for those looking to grasp the basics and delve deeper.

—Sergey Zakharchenko, Senior Engineer, Doxel

This book offers a comprehensive guide to navigating the realm of data for machine learning. From curating trainable data across commonly used formats to refining it for production, it's a good read to expand your data acumen.

—Satyarth Praveen, Computer Systems Engineer,
Lawrence Berkeley National Lab

Data has long been considered one of the most valuable assets for any company, but there's a difference between raw data and actionable data. ML introduces a radical new way in how data can be utilized, but it also introduces new complexities. The ability to use that data for training primarily depends on how well it is prepared and organized, and this book serves as a helpful guide.

—Vladimir Mandic, Machine Learning Consultant, and former
CTO, Dell/EMC Data Protection Software and Cloud divisions

Anthony Sarkis carefully dissects and presents an often overlooked part of the AI pipeline with a novel conceptualization of it as a separate paradigm. A must-read for everyone who is interested in more than doing ML in a vacuum.

—Ihor Markevych, Lead AI Engineer, Cognistx

Having good training data is crucial for training robust machine learning systems. This beginner-friendly book gives a great overview of how to get good datasets from day one.

—Igor Susmelj, Cofounder, Lightly

Anthony Sarkis provides readers with an excellent and approachable resource on the role of training data in modern machine learning systems. It is a must-read for anyone interested in data annotation projects.

—*Zygmunt Lenyk, Senior Research Engineer, Five AI Limited*

Anthony brilliantly unravels the intricate details and processes of machine learning data preparation. This book offers a clear roadmap for organizations navigating the often underestimated terrain of training data.

—*Pablo Estrada, Senior Software Engineer, Diffgram*

The data labeling process sounds obvious until you experience mistakes firsthand, including caveats, gotchas, unplanned complexity, and trade-offs. Thanks to this book's in-depth examples, you won't need to, because it provides a thorough 360-degree view of how to generate high-quality training data and jump-start new projects while knowing the dangers to look out for.

—*Anirudh Koul,*
Head of Machine Learning Data Science, Pinterest

Training Data for Machine Learning is essential reading for the CSuite as well as the AI engineer. Sarkis's writing is insightful and wise as he presents impactful information that will save your business time and money. Every large language model practitioner will significantly improve their LLM by using this excellent resource.

—*Neal Linson, Chief Data and Analytics Officer,*
InCite Logix, LLM superstar

This book throws light on an important aspect of machine learning–the data being used for model training. Readers can find resources on ML models and algorithms but a model is only good as the data is being trained on. This book completes the picture for data-driven model training, highlighting the best practices in industry through several case studies.

—*Prabhav Agrawal, Machine Learning Engineer, Meta*

A refreshing take on machine learning literature, this book goes beyond the typical ML algorithms and delves into the often overlooked, yet crucial, world of training data. It offers an invaluable foundation for managing data, making it indispensable for anyone in the ML landscape.

—*Tarun Narayanan V., AI Researcher,*
NASA Frontier Development Lab

Training Data for Machine Learning
Human Supervision from Annotation to Data Science

Anthony Sarkis

Beijing • Boston • Farnham • Sebastopol • Tokyo

Training Data for Machine Learning

by Anthony Sarkis

Published by O'Reilly Media, Inc., 1005 Gravenstein Highway North, Sebastopol, CA 95472.

O'Reilly books may be purchased for educational, business, or sales promotional use. Online editions are also available for most titles (*http://oreilly.com*). For more information, contact our corporate/institutional sales department: 800-998-9938 or *corporate@oreilly.com*.

Acquisitions Editor: Aaron Black	**Indexer:** nSight, Inc.
Development Editor: Jill Leonard	**Interior Designer:** David Futato
Production Editor: Elizabeth Faerm	**Cover Designer:** Karen Montgomery
Copyeditor: Liz Wheeler	**Illustrator:** Kate Dullea
Proofreader: Dwight Ramsey	

November 2023:   First Edition

Revision History for the First Edition
2023-11-08:   First Release

See *http://oreilly.com/catalog/errata.csp?isbn=9781492094524* for release details.

This work is part of a collaboration between O'Reilly and Kili Technology. See our statement of editorial independence (*https://oreil.ly/editorial-independence*).

978-1-492-09452-4

[LSI]

Table of Contents

Preface

Any sufficiently advanced technology is indistinguishable from magic.
—Arthur C. Clarke

Your work, career, or daily life may already be, or about to be, impacted by artificial intelligence (AI). This book will help build and improve your understanding of the concepts and mechanics behind a key part of AI called training data.

Will your life really be affected? Here's a test. Are you in tech, working on a software product? Does your work, or your company's product, have any form of repetitive tasks? Things you or users of your product do on a regular cycle? If you answered yes to any of these questions, AI and machine learning (ML) have the potential to take on more of the workload, enabling you or your users to focus on higher-level work, and will therefore impact you. If you want to better align with this new AI wave, this book will reveal many of the nuts and bolts that make AI actually work in practice. It will help accelerate your success in your current job and prepare you for new AI-centric roles.

Speaking of jobs, you know how the first few days or weeks on a new job are—stressful, crazy, unpredictable? Then all of the sudden, the work, all the day-to-day stuff, snaps into place and makes sense? What was once unimaginable becomes ordinary because you learned how to fit in, how to adapt. Over a relatively short period of time you go from spilling coffee on the boss's shirt to being a productive part of the system.

AI works in a similar way. The difference is that the boss of the AI is you! You are responsible for the initial and ongoing training of the AI. Like a new team member, when the AI is first training, the results are unpredictable. Over time, as you train and supervise it more, it gets better. In fact, it happens so quickly, that the assumptions around what is automatable and what is not get turned upside down. This high-level supervision is required for all AI systems, from self-driving cars to agricultural weed detection, medical diagnostics, security, sports analytics, and more.

I will pull back the curtain on this most fundamental aspect of AI: mapping human meaning into an AI-readable form, otherwise known as training data. This matters for everything from generative AI to fully supervised systems. I will help you understand the many representations and concepts around training data. We'll cover how it works in practice, including operations, tooling, automation, and system design. And we will put it all together with practical case studies and tips.

Your knowledge is the magic that makes AI work. AI enables you to expand your reach. To do more creative work. To multiply the effectiveness of your knowledge. And if you can learn how to train AI, then you will be the beneficiary.

Who Should Read This Book?

This book is a foundational overview of training data. It's ideally suited to those who are totally new, or just getting started, with training data.

For intermediate practitioners, the later chapters provide unique value and insights that can't be found anywhere else; in a nutshell, insider knowledge. I will highlight specific areas of interest for subject matter experts, workflow managers, directors of training data, data engineers, and data scientists.

Computer science (CS) knowledge is not required. Knowing CS, machine learning, or data science will make more sections of the book accessible. I strive to make this book maximally accessible to data annotators, including subject matter experts, because they play a key part in training data, including supervising the system.

For the Technical Professional and Engineer

You may have been looking to ship or improve a system, or you've seen a new AI capability being demonstrated and are looking to apply it to your domain. This book will guide you through these processes, as well as addressing more detailed questions, such as what media types you should use, how to wire up the system, and what automations matter.

There are many approaches, and this book aims to provide you with a balanced coverage, highlight trade-offs, and be a primary reference for your training data needs. There are so many new concepts that keeping informed may sometimes feel like reading the docs. Whenever possible, I have tried my best to focus the book's style on the "in-between," coffee-shop type talk that doesn't exist on public docs.

If you are an expert, then this book can act as a reference, refresher, and easy way to convey core concepts to new people on your team. Already have some knowledge in this area and want a gut check on completeness? This book will expand your toolkit of approaches and provide new perspectives on common ideas. And if you are completely new, this is your best resource to get started.

For the Manager and Director

Simply put, this book has content you can't get anywhere else. The uniqueness and density of the context this book adds is novel and will help you and your team unlock insights, potentially getting you months or even years ahead.

Further, significant sections of this book are devoted to people and processes. Training data presents novel human–computer interaction concepts and involves levels of cross-disciplinary interaction that will provide valuable new insights, ensuring your success in this exciting field of AI.

You may especially be interested in Chapter 6, "Theories, Concepts, and Maintenance", Chapter 7, "AI Transformation and Use Cases", and Chapter 9, "Case Studies and Stories". The rest of the chapters will help you feel comfortable enough with the details to be able to recognize successes and failures; this will help with course corrections.

For the Subject Matter Expert and Data Annotation Specialist

Annotators are one of the absolute most critical roles for the daily production of training data. The 2020 World Economic Forum report says the top three job roles showing an increase in demand all involve data analysis and AI.[1] Knowing how to work with training data is a valuable skill to add to your existing skills, and also a new career opportunity in its own right.

It is increasingly common for employers to ask all employees to understand the basics around AI, and often even training data. For example, a large auto employer states on their data annotator job description that the applicant must: "Know how the labels are used by our learning algorithms so as to make more judgment calls on difficult edge cases."[2] No matter your industry or background, you have a huge opportunity to expand the reach of your knowledge and the productivity of your company by grounding your and your team's knowledge in training data.

While anyone can supervise areas they are knowledgeable on, subject matter experts (SMEs) such as doctors, lawyers, and engineers can be especially valuable. SMEs can both supervise AI directly and provide detailed instructions and training toward more cost-effective resources. If you're an SME, it's vital to read this book, even more closely, to understand how your work fits into the AI picture, what knobs and levers are available for you to use, and how to set up processes for other people to follow.

---

1 "The Future of Jobs Report 2020" (*https://oreil.ly/m6uXd*), *World Economic Forum*, October 2020 (p. 30, Figure 22).

2 "Data Annotation Specialist", Tesla website, accessed November 5, 2020.

This book will also help provide insight into tested mechanics, such as a concept called the *schema*, in addition to standard material like detailed instructions. By reading this book, you will gain a deep understanding of everything you need to create and maintain effective AI systems through training data.

For the Data Scientist

As a data scientist, you have an important role to play as an advisor to others: to help them understand how the data will actually be used. Even the most advanced and integrated AutoML systems usually need somebody to interpret and understand the meaning of their output, and to be able to debug them when something goes wrong. This book will help you better engage with your diverse annotation and technical partners.

Any data can be trained on or considered training data. Like many terms ("apple" the fruit versus "Apple" the company), training data has multiple meanings. This book focuses on supervised training data, meaning a human is involved in enriching the data directly. While the details of annotation may not always be relevant to your day-to-day work, a broader understanding can help ensure the end result is the best possible.

To set expectations, this book focuses on modern training data, and specifically supervised systems where a human plays at least some role. Even in the generative AI context, which is often thought of as unsupervised, human alignment plays a key role. While the boundaries or usefulness of concepts around supervised, self-supervised, semi-supervised, unsupervised, you-name-it-supervised remain in flux, it appears clear that many practical use cases are achievable with some degree of supervision, and that supervision in some form will likely be here for a long time to come.

While reading, here are some themes to consider. How can you more deeply engage with your annotation and technical partners? How can you be involved in the dataset processes, including creation and maintenance? How can you help align your modeling needs to the schema and vice versa? How can you help ensure the training data will be the best it can be for your models? If there's one takeaway from this book, I hope it's that you see "data annotation" in a new light, as its own technology area, called training data.

Why I Wrote This Book

Throughout my journey with Diffgram, I have noticed a vast gulf between those who "got it" and those who didn't. Often, it felt like I was watching someone trying to learn multiplication before they knew the number system existed. The basic foundations of training data were missing (and worse, they often didn't realize they were missing!).

Initially, I just started to write short articles—relatively brief, mostly only a few pages, focused on a narrow topic. These helped "spot treat" knowledge gaps. It was me sharing things in my little niche, things I happened to know. But it still felt like large sections were missing. I needed to write something more holistic. A book was the logical next step. But, who was I to write it?

When I started writing this I had a lot of doubts. I had been working in the area for about three years already, but I still felt like some of the material I planned to write was an "aspirational" goal—not just summarizing what I already knew. As I write this section today, reflecting on now five years, I still feel like I have barely scratched the surface of this area.

However, at this point, I had to look back and realize that there were very few other people I knew of who had maintained a comparable level of in-depth technical understanding as their businesses had scaled. This meant I was on a short list of people who have a particular set of characteristics: a deep technical understanding of this area, knowledge of the history of its progress, the ability to explain these topics in non-engineer terms, and the desire to take the time to record and share that knowledge with others.

I really believe training data is one of the most major conceptual shifts in technology to have appeared in a long time. Supervised training data crosscuts every industry and virtually every product. Over the next few decades, I believe it will shape our lives in ways we can barely imagine today. I hope this book helps you on your journey.

How This Book Is Organized

First, I introduce what you can do with training data, opportunities in working with training data, why training data matters, and training data in the wild (Chapter 1, "Training Data Introduction"). Real-world projects require training data tools, and it helps to ground concepts when you are actually able to work with them. To help you get started (Chapter 2, "Getting Up and Running"), we'll offer you just that—a framework to go ahead and get to work.

Once you have the high-level concepts and some tooling, it'll be time to talk about the schema—the paradigm for encoding all of your commercial knowledge. The schema is one of the most important concepts in training data, so this detailed treatment (Chapter 3, "Schema") will really help build that understanding. Next are data engineering (Chapter 4, "Data Engineering") and workflow (Chapter 5, "Workflow"), key engineering concepts to help you get your system up and into production.

We then transition into concepts and theories (Chapter 6, "Theories, Concepts, and Maintenance"), AI transformation (Chapter 7, "AI Transformation and Use Cases"), and automation (Chapter 8, "Automation"), and conclude with real-world case studies (Chapter 9, "Case Studies and Stories").

Themes

This book is divided into three major themes, as follows.

The Basics and Getting Started

Learn why training data is important and what it is. Get baseline terminologies, concepts, and types of representations down pat. I frame the context, starting with similarities and differences between supervised and classic approaches to ML. I then unpack all the aspects around abstractions, people, process, and more. This is the foundational baseline.

Concepts and Theories

Here, we get more specific, looking at system and user operations and popular automation approaches. Here we move slightly away from the foundational and expand toward varying opinions.

Putting It All Together

Taking both the foundational and theory needs in mind, we explore specific implementations. We further expand the trends to cover cutting-edge research topics and directions.

A small note on terms: throughout this book, you'll occasionally see the terms training data and AI data used synonymously. AI data is a broad term referring to any type of data used by AI. All training data is AI data too.

Often, I will use analogies to help make the content more accessible and memorable. I will purposely avoid technical jargon unless its inclusion is critical. If you're an expert, please disregard anything you're already familiar with; for the non-expert, please consider that a lot of the technical details are just that—details. The details add to the understanding but are not required for it.

I aim to stay as focused on supervised training data as possible. This includes brief forays into deep learning and ML knowledge, but generally, that's out of scope. Training data is a general-purpose concept across industries, applying equally well to many. The majority of the concepts presented apply equally well to multiple domains.

Despite my firsthand experience with the evolution of ML and AI, this is not a history book; I will only cover the history enough so as to ground the current topics.

Software built around training data introduces a variety of assumptions and constraints. I attempt to unearth hidden assumptions, and highlight concepts commonly known in select circles, but new to most everyone else.

Conventions Used in This Book

The following typographical conventions are used in this book:

Italic
: Indicates new terms, URLs, email addresses, filenames, and file extensions.

`Constant width`
: Used for program listings, as well as within paragraphs to refer to program elements such as variable or function names, databases, data types, environment variables, statements, and keywords.

`Constant width bold`
: Shows commands or other text that should be typed literally by the user.

`Constant width italic`
: Shows text that should be replaced with user-supplied values or by values determined by context.

This element signifies a general note.

O'Reilly Online Learning

 For more than 40 years, *O'Reilly Media* has provided technology and business training, knowledge, and insight to help companies succeed.

Our unique network of experts and innovators share their knowledge and expertise through books, articles, and our online learning platform. O'Reilly's online learning platform gives you on-demand access to live training courses, in-depth learning paths, interactive coding environments, and a vast collection of text and video from O'Reilly and 200+ other publishers. For more information, visit *https://oreilly.com*.

How to Contact Us

Please address comments and questions concerning this book to the publisher:

O'Reilly Media, Inc.
1005 Gravenstein Highway North
Sebastopol, CA 95472
800-889-8969 (in the United States or Canada)
707-829-7019 (international or local)
707-829-0104 (fax)
support@oreilly.com
https://www.oreilly.com/about/contact.html

We have a web page for this book, where we list errata, examples, and any additional information. You can access this page at *https://oreil.ly/training-data-for-ml*.

For news and information about our books and courses, visit *https://oreilly.com*.

Find us on LinkedIn: *https://linkedin.com/company/oreilly-media*.

Follow us on Twitter: *https://twitter.com/oreillymedia*.

Watch us on YouTube: *https://youtube.com/oreillymedia*.

Acknowledgments

I would like to thank Pablo Estrada, Vitalii Bulyzhyn, Sergey Zakharchenko and Francesco Virga for their work at Diffgram that enabled me to write this book and feedback on early versions. I'd like to thank mentors Vladimir Mandic and Neal Linson for their encouragement and advice. I'd also like to thank Xue Hai Fang, Luba Kozak, Nathan Muchowski, Shivangini Chaudhary, Tanya Walker, and Michael Sarkis for their resolute support.

I also would like to thank the reviewers who provided valuable feedback, including Igor Susmelj, Tarun Narayanan, Ajay Krishnan, Satyarth Praveen, Prabhav Agrawal, Kunal Khadilkar, Zygmunt Lenyk, Giovanni Alzetta, and Ihor Markevych.

And of course the excellent staff at O'Reilly, especially Jill Leonard who stuck with me and this book throughout its development and Aaron Black for his timely wisdom.

CHAPTER 1

Training Data Introduction

Data is all around us—videos, images, text, documents, as well as geospatial, multi-dimensional data, and more. Yet, in its raw form, this data is of little use to supervised machine learning (ML) and artificial intelligence (AI). How do we make use of this data? How do we record our intelligence so it can be reproduced through ML and AI? The answer is the art of training data—the discipline of making raw data useful.

In this book you will learn:

- All-new training data (AI data) concepts
- The day-to-day practice of training data
- How to improve training data efficiency
- How to transform your team to be more AI/ML-centric
- Real-world case studies

Before we can cover some of these concepts, we first have to understand the foundations, which this chapter will unpack.

Training data is about molding, reforming, shaping, and digesting raw data into new forms: creating new meaning out of raw data to solve problems. These acts of creation and destruction sit at the intersection of subject matter expertise, business needs, and technical requirements. It's a diverse set of activities that crosscut multiple domains.

At the heart of these activities is annotation. Annotation produces structured data that is ready to be consumed by a machine learning model. Without annotation, raw data is considered to be unstructured, usually less valuable, and often not usable for supervised learning. That's why training data is required for modern machine learning use cases including computer vision, natural language processing, and speech recognition.

To cement this idea in an example, let's consider annotation in detail. When we annotate data, we are capturing human knowledge. Typically, this process looks as follows: a piece of media such as an image, text, video, 3D design, or audio, is presented along with a set of predefined options (labels). A human reviews the media and determines the most appropriate answers, for example, declaring a region of an image to be "good" or "bad." This label provides the context needed to apply machine learning concepts (Figure 1-1).

But how did we get there? How did we get to the point that the right media element, with the right predefined set of options, was shown to the right person at the right time? There are many concepts that lead up to and follow the moment where that annotation, or knowledge capture, actually happens. Collectively, all of these concepts are the art of training data.

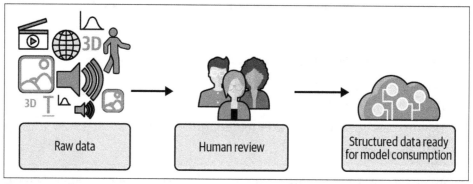

Figure 1-1. The training data process

In this chapter, we'll introduce what training data is, why it matters, and dive into many key concepts that will form the base for the rest of the book.

Training Data Intents

The purpose of training data varies across different use cases, problems, and scenarios. Let's explore some of the most common questions, like what can you do with training data? What is it most concerned with? What are people aiming to achieve with training data?

What Can You Do With Training Data?

Training data is the foundation of AI/ML systems—the underpinning that makes these systems work.

With training data, you can build and maintain modern ML systems, such as ones that create next-generation automations, improve existing products, and even create all-new products.

In order to be most useful, the raw data needs to be upgraded and structured in a way that is consumable by ML programs. With training data, you are creating and maintaining the required new data and structures, like annotations and schemas, to make the raw data useful. Through this creation and maintenance process, you will have great training data, and you will be on the path toward a great overall solution.

In practice, common use cases center around a few key needs:

- Improving an existing product (e.g., performance), even if ML is not currently a part of it
- Production of a new product, including systems that run in a limited or "one-off" fashion
- Research and development

Training data transcends all parts of ML programs:

- Training a model? It requires training data.
- Want to improve performance? It requires higher-quality, different, or a higher volume of training data.
- Made a prediction? That's future training data that was just generated.

Training data comes up before you can run an ML program; it comes up during running in terms of output and results, and even later in analysis and maintenance. Further, training data concerns tend to be long-lived. For example, after getting a model up and running, maintaining the training data is an important part of maintaining a model. While, in research environments, a single training dataset may be unchanged (e.g., ImageNet), in industry, training data is extremely dynamic and changes often. This dynamic nature puts more and more significance on having a great understanding of training data.

The creation and maintenance of novel data is a primary concern of this book. A dataset, at a moment in time, is an output of the complex processes of training data. For example, a Train/Test/Val split is a derivative of an original, novel set. And that novel set itself is simply a snapshot, a single view into larger training data processes. Similarly to how a programmer may decide to print or log a variable, the variable printed is just the output; it doesn't explain the complex set of functions that were required to get the desired value. A goal of this book is to explain the complex processes behind getting usable datasets.

Annotation, the act of humans directly annotating samples, is the "highest" part of training data. By highest, I mean that human annotation works on top of the collection of existing data (e.g., from BLOB storages, existing databases, metadata, websites).[1] Human annotation is also the overriding truth on top of automation concepts like pre-labeling and other processes that generate new data like predictions and tags. These combinations of "high-level" human work, existing data, and machine work form a core of the much broader concepts of training data outlined later in this chapter.

What Is Training Data Most Concerned With?

This book covers a variety of people, organizational, and technical concerns. We'll walk through each of these concepts in detail in a moment, but before we do, let's think about areas training data is focused on.

For example, how does the schema, which is a map between your annotations and their meaning for your use case, accurately represent the problem? How do you ensure raw data is collected and used in a way relevant to the problem? How are human validation, monitoring, controls, and correction applied?

How do you repeatedly achieve and maintain acceptable degrees of quality when there is such a large human component? How does it integrate with other technologies, including data sources and your application?

To help organize this, you can broadly divide the overall concept of training data into the following topics: schema, raw data, quality, integrations, and the human role. Next, I'll take a deeper look at each of those topics.

Schema

A schema is formed through labels, attributes, spatial representations, and relations to external data. Annotators use the schema while making annotations. Schemas are the backbone of your AI and central in every aspect of training data.

Conceptually, a schema is the map between human input and meaning for your use case. It defines what the ML program is capable of outputting. It's the vital link, it's what binds together everyone's hard work. So to state the obvious, it's important.

---

1 In most cases, that existing data is thought of as a "sample," even if it was created by a human at some point prior.

A good schema is useful and relevant to your specific need. It's usually best to create a new, custom schema, and then keep iterating on it for your specific cases. It's normal to draw on domain-specific databases for inspirations, or to fill in certain levels of detail, but be sure that's done in the context of guidance for a new, novel, schema. Don't expect an existing schema from another context to work for ML programs without further updates.

So, why is it important to design it according to your specific needs, and not some predefined set?

First, the schema is for both human annotation and ML use. An existing domain-specific schema may be designed for human use in a different context or for machine use in a classic, non-ML context. This is one of those cases where two things might seem to produce similar output, but the outcome is actually formed in totally different ways. For example, two different math functions might both output the same value, but run on completely different logic. The output of the schema may appear similar, but the differences are important to make it friendly to annotation and ML use.

Second, if the schema is not useful, then even great model predictions are not useful. Failure with schema design likely will cascade to failure of the overall system. The context here is that ML programs can usually only make predictions based on what is included in the schema.[2] It's rare that an ML program will produce relevant results that are better than the original schema. It's also rare that it will predict something that a human, or group of humans, looking at the same raw data could not also predict.

It is common to see schemas that have questionable value. So, it's really worth stopping and thinking "If we automatically got the data labeled with this schema, would it actually be useful to us?" and "Can a human looking at the raw data reasonably choose something from the schema that fits it?"

In the first few chapters, we will cover the technical aspects of schemas, and we will come back to schema concerns through practical examples later in the book.

Raw data

Raw data is any form of Binary Large Object (BLOB) data or pre-structured data that is treated as a single sample for purposes of annotation. Examples include videos, images, text, documents, and geospatial and multi-dimensional data. When we think about raw data as part of training data, the most important thing is that the raw data is collected and used in a way relevant to the schema.

---

2 Without further deductions outside our scope of concern.

To illustrate the idea of the relevance of raw data to a schema, let's consider the difference between hearing a sports game on the radio, seeing it on TV, or being at the game in person. It's the same event regardless of the medium, but you receive a very different amount of data in each context. The context of the raw data collection, via TV, radio, or in person, frames the potential of the raw data. So, for example, if you were trying to determine possession of the ball automatically, the visual raw data will likely be a better fit then the radio raw data.

Compared to software, we humans are good at automatically making contextual correlations and working with noisy data. We make many assumptions, often drawing on data sources not present in the moment to our senses. This ability to understand the context above the directly sensed sights, sounds, etc. makes it difficult to remember that software is more limited here.

Software only has the context that is programmed into it, be it through data or lines of code. This means the real challenge with raw data is overcoming our human assumptions around context to make the right data available.

So how do you do that? One of the more successful ways is to start with the schema, and then map ideas of raw data collection to that. It can be visualized as a chain of problem -> schema -> raw data. The schema's requirements are always defined by the problem or product. That way there is always this easy check of "Given the schema, and the raw data, can a human make a reasonable judgment?"

Centering around the schema also encourages thinking about new methods of data collection, instead of limiting ourselves to existing or easiest-to-reach data collection methods. Over time, the schema and raw data can be jointly iterated on; this is just to get started. Another way to relate the schema to the product is to consider the schema as representing the product. So to use the cliché of "product market fit," this is "product data fit."

To put the above abstractions into more concrete terms, we'll discuss some common issues that arise in industry. Differences between data used during development and production is one of the most common sources of errors. It is common because it is somewhat unavoidable. That's why being able to get to some level of "real" data early in the iteration process is crucial. You have to expect that production data will be different, and plan for it as part of your overall data collection strategy.

The data program can *only* see the raw data and the annotations—only what is given to it. If a human annotator is relying on knowledge outside of what can be understood from the sample presented, it's unlikely the data program will have that context, and it will fail. We must remember that all needed context must be present, either in the data or lines of code of the program.

To recap:

- The raw data needs to be relevant to the schema.
- The raw data should be as similar to production data as possible.
- The raw data should have all the context needed in the sample itself.

Annotations

Each annotation is a single example of something specified in the schema. Imagine two cliffs with an open space in the middle, the left representing the schema, and the right a single raw data file. An annotation is the concrete bridge between the schema and raw data, as shown in Figure 1-2.

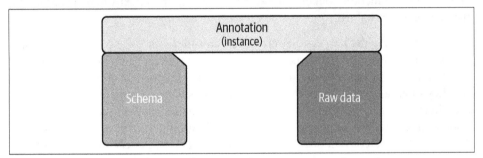

Figure 1-2. Relationships among schema, single annotation, and raw data

While the schema is "abstract," meaning that it is referenced and reused between multiple annotations, each annotation has the actual specific values that fill in the answers to the questions in the schema.

Annotations are usually the most numerous form of data in a training data system because each file often has tens, or even hundreds, of annotations. An annotation is also known as an "instance" because it's a single instance of something in the schema.

More technically, each annotation instance usually contains a key to relate it to a label or attribute within a schema and a file or child file representing the raw data. In practical terms, each file will usually contain a list of instances.

Quality

Training data quality naturally occurs on a spectrum. What is acceptable in one context may not be in another.

So what are the biggest factors that go into training data quality? Well, we already talked about two of them: schema and raw data. For example:

- A bad schema may cause more quality issues than bad annotators.
- If the concept is not clear in the raw data sample, it's unlikely it will be clear to the ML program.

Often, annotation quality is the next biggest item. Annotation quality is important, but perhaps not in the ways you may expect. Specifically, people tend to think of annotation quality as "was it annotated right?" But "right" is often out of scope. To understand how the "right" answer is often out of scope, let's imagine we are annotating traffic lights, and the light in the sample you are presented with is off (e.g., a power failure) and your only options from the schema are variations on an active traffic light. Clearly, either the schema needs to be updated to include an "off" traffic light, or our production system will never be usable in a context where a traffic light may have a power failure.

To move into a slightly harder-to-control case, consider that if the traffic light is really far away or at an odd angle, that will also limit the worker's ability to annotate it properly. Often, these cases sound like they should be easily manageable, but in practice they often aren't. So more generally, real issues with annotation quality tend to circle back to issues with the schema and raw data. Annotators surface problems with schemas and data in the course of their work. High-quality annotation is as much about the effective communication of these issues rather than exclusively annotating "correctly."

I can't emphasize enough that schema and raw data deserve a lot of attention. However, annotating correctly does still matter, and one of the approaches is to have multiple people look at the same sample. This is often costly, and someone must interpret the meaning of the multiple opinions on the same sample, adding further cost. For an industry-usable case, where the schema has a reasonable degree of complexity, the meta-analysis of the opinions is a further time sink.

Think of a crowd of people watching a sports game instant replay. Imagine trying to statistically sample their opinions to get a "proof" of what is "more right." Instead of this, we have a referee who individually reviews the situation and makes a determination. The referee may not be "right" but, for better or worse, the social norm is to have the referee (or a similar process) make the call.

Similarly, often a more cost-effective approach is used. A percent of the data is sampled randomly for a review loop, and annotators raise issues with the schema and raw data fit, as they occur. This review loop and quality assurance processes will be discussed in more depth later.

If the review method fails, and it seems you would still need multiple people to annotate the same data in order to ensure high quality, you probably have a bad product data fit, and you need to change the schema or raw data collection to fix it.

Zooming out from schema, raw data, and annotation, the other big aspects of quality are the maintenance of the data and the integration points with ML programs. Quality includes cost considerations, expected use, and expected failure rates.

To recap, quality is first and foremost formed by the schema and raw data, then by the annotators and associated processes, and rounded out by maintenance and integration.

Integrations

Much time and energy are often focused on "training the model." However, because training a model is a primarily technical data-science-focused concept, it can lead us to underemphasize other important aspects of using the technology effectively.

What about maintenance of the training data? What about ML programs that output useful training data results, such as sampling, finding errors, reducing workload, etc., that are not involved with training a model? How about the integration with the application that the results of the model or ML subprogram will be used in? What about tech that tests and monitors datasets? The hardware? Human notifications? How the technology is packaged into other tech?

Training the model is just one component. To successfully build an ML program, a data-driven program, we need to think about how all the technology components work together. And to hit the ground running, we need to be aware of the growing training data ecosystem. Integration with data science is multifaceted, it's not just about some final "output" of annotations. It's about the ongoing human control, maintenance, schema, validation, lifecycle, security, etc. A batch of outputted annotations is like the result of a single SQL query, it's a single, limited view into a complex database.

A few key aspects to remember about working with integrations:

- The training data is only useful if it can be consumed by something, usually within a larger program.
- Integration with data science has many touch points and requires big-picture thinking.
- Getting a model trained is only a small part of the overall ecosystem.

The human role

Humans affect data programs by controlling the training data. This includes determining the aspects we have discussed so far, the schema, raw data, quality, and integrations with other systems. And of course, people are involved with annotation itself, when humans look at each individual sample.

This control is exercised at many stages, and by many people, from establishing initial training data to performing human evaluations of data science outputs and validating data science results. This large volume of people being involved is very different from classic ML.

We have new metrics, like how many samples were accepted, how long is spent on each task, the lifecycles of datasets, the fidelity of raw data, what the distribution of the schema looks like, etc. These aspects may overlap with data science terms, like class distribution, but are worth thinking of as separate concepts. For example, model metrics are based on the ground truth of the training data, so if the data is wrong, the metrics are wrong. And as discussed in "Quality Assurance Automation" on page 259, metrics around something like annotator agreement can miss larger points of schema and raw data issues.

Human oversight is about so much more than just quantitative metrics. It's about qualitative understanding. Human observation, human understanding of the schema, raw data, individual samples, etc., are of great importance. This qualitative view extends into business and use case concepts. Further, these validations and controls quickly extend from being easily defined, to more of an art form, acts of creation. This is not to mention the complicated political and social expectations that can arise around system performances and output.

Working with training data is an opportunity to create: to capture human intelligence and insights in novel ways; to frame problems in a new training data context; to create new schemas, to collect new raw data, and use other training data–specific methods.

This creation, this control, it's all new. While we have established patterns for various types of human–computer interaction, there is much less established for human–ML program interactions—for human supervision, a data-driven system, where the humans can directly correct and program the data.

For example, we expect an average office worker to know how to use word processing, but we don't expect them to use video editing tools. Training data requires subject matter experts. So in the same way a doctor must know how to use a computer for common tasks today, they must now learn how to use standard annotation patterns. As human-controlled data-driven programs emerge and become more common, these interactions will continue to increase in importance and variance.

Training Data Opportunities

Now that we understand many of the fundamentals, let's frame some opportunities. If you're considering adding training data to your ML/AI program, some questions you may want to ask are:

- What are the best practices?
- Are we doing this the "right" way?
- How can my team work more efficiently with training data?
- What business opportunities can training data–centric projects unlock?
- Can I turn an existing work process, like an existing quality assurance pipeline, into training data? What if all of my training data could be in one place instead of shuffling data from A to B to C? How can I be more proficient with training data tools?

Broadly, a business can:

- Increase revenue by shipping new AI/ML data products.
- Maintain existing revenue by improving the performance of an existing product through AI/ML data.
- Reduce security risks—reduce risks and costs from AI/ML data exposure and loss.
- Improve productivity by moving employee work further up the automation food chain. For example, by continuously learning from data, you can create your AI/ML data engine.

All of these elements can lead to transformations through an organization, which I'll cover next.

Business Transformation

Your team and company's mindset around training data is important. I'll provide more detail in Chapter 7, but for now, here are some important ways to start thinking about this:

- Start viewing all existing routine work at the company as an opportunity to create training data.
- Realize that work not captured in a training data system is lost.
- Begin shifting annotation to be part of every frontline worker's day.

- Define your organizational leadership structures to better support training data efforts.

- Manage your training data processes at scale. What works for an individual data scientist might be very different from what works for a team, and different still for a corporation with multiple teams.

In order to accomplish all of this, it's important to implement strong training data practices within your team and organization. To do this, you need to create a training data–centric mindset at your company. This can be complex and may take time, but it's worth the investment.

To do this, involve subject matter experts in your project planning discussions. They'll bring valuable insights that will save your team time downstream. It's also important to use tools to maintain abstractions and integrations for raw data collection, ingress, and egress. You'll need new libraries for specific training data purposes so you can build on existing research. Having the proper tools and systems in place will help your team perform with a data-centric mindset. And finally, make sure you and your teams are reporting and describing training data. Understanding what was done, why it was done, and what the outcomes were will inform future projects.

All of this may sound daunting now, so let's break things down a step further. When you first get started with training data, you'll be learning new training data–specific concepts that will lead to mindset shifts. For example, adding new data and annotations will become part of your routine workflows. You'll be more informed as you get initial datasets, schemas, and other configurations set up. This book will help you become more familiar with new tools, new APIs, new SDKs, and more, enabling you to integrate training data tools into your workflow.

Training Data Efficiency

Efficiency in training data is a function of many parts. We'll explore this in greater detail in the chapters to come, but for now, consider these questions:

- How can we create and maintain better schemas?

- How can we better capture and maintain raw data?

- How can we annotate more efficiently?

- How can we reduce the relevant sample counts so there is less to annotate in the first place?

- How can we get people up to speed on new tools?

- How can we make this work with our application? What are the integration points?

As with most processes, there are a lot of areas to improve efficiency, and this book will show you how sound training data practices can help.

Tooling Proficiency

New tools, like Diffgram, HumanSignal, and more now offer many ways to help realize your training data goals. As these tools grow in complexity, being able to master them becomes more important. You may have picked up this book looking for a broad overview, or to optimize specific pain points. Chapter 2 will discuss tools and trade-offs.

Process Improvement Opportunities

Consider a few common areas people want to improve, such as:

- Annotation quality being poor, too costly, too manual, too error prone
- Duplicate work
- Subject matter expert labor cost too high
- Too much routine or tedious work
- Nearly impossible to get enough of the original raw data
- Raw data volume clearly exceeding any reasonable ability to manually look at it

You may want a broader business transformation, to learn new tools, or to optimize a specific project or process. The question is, naturally, what's the next best step for you to take, and why should you take it? To help you answer that, let's now talk about why training data matters.

Why Training Data Matters

In this section, I'll cover why training data is important for your organization, and why a strong training data practice is essential. These are central themes throughout the book, and you'll see them come up again in the future.

First, training data determines what your AI program, your system, can do. Without training data, there is no system. With training data, the opportunities are only bounded by your imagination! Sorta. Well, okay, in practice, there's still budget, resources such as hardware, and team expertise. But theoretically, anything that you can form into a schema and record raw data for, the system can repeat. Conceptually, the model can learn anything. Meaning, the intelligence and ability of the system depends on the quality of the schema, and the volume and variety of data you can teach it. In practice, effective training data gives you a key edge when all else—budget, resources, etc.—are equal.

Second, training data work is upstream, before data science work. This means data science is dependent on training data. Errors in training data flow down to data science. Or to use the cliché—garbage in, garbage out. Figure 1-3 walks through what this data flow looks like in practice.

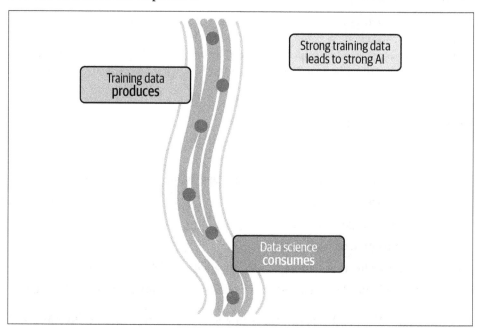

Figure 1-3. Conceptual positions of training data and data science

Third, the art of training data represents a shift in thinking about how to build AI systems. Instead of over-focusing on improving mathematical algorithms, in parallel with them, we continue to optimize the training data to better match our needs. This is the heart of the AI transformation taking place, and the core of modern automation. For the first time, knowledge work is now being automated.

ML Applications Are Becoming Mainstream

In 2005, a university team used a training data–based[3] approach to engineer a vehicle, Stanley, that could drive autonomously on an off-road 175-mile-long desert course, winning the Defense Advanced Research Projects Agency (DARPA) grand challenge

---

3 From "Stanley_(vehicle)" (*https://oreil.ly/SLAGV*), *Wikipedia*, accessed on September 8, 2023: "Stanley was characterized by a machine learning based approach to obstacle detection. To correct a common error made by Stanley early in development, the Stanford Racing Team created a log of 'human reactions and decisions' and fed the data into a learning algorithm tied to the vehicle's controls; this action served to greatly reduce Stanley's errors. The computer log of humans driving also made Stanley more accurate in detecting shadows, a problem that had caused many of the vehicle failures in the 2004 DARPA Grand Challenge."

(*https://oreil.ly/byk0x*). About 15 years later, in October 2020, an automotive company released a controversial *Full Self-Driving (FSD)* technology (*https://oreil.ly/lvmz9*) in public, ushering in a new era of consumer awareness. In 2021, data labeling concerns started getting mentioned on earnings calls. In other words, the mainstream is starting to get exposed to training data.

This commercialization goes beyond headlines of AI research results. In the last few years, we have seen the demands placed on technology increase dramatically. We expect to be able to speak to software and be understood, to automatically get good recommendations and personalized content. Big tech companies, startups, and businesses alike are increasingly turning to AI to address this explosion in use case combinations.

AI knowledge, tooling, and best practices rapidly expand. What used to be the exclusive domain of a few is now becoming common knowledge, and pre-built API calls. We are at the transition phase, going from R&D demos to the early stages of real-world industry use cases.

Expectations around automation are being redefined. Cruise control, to a new car buyer, has gone from just "maintain constant speed" to include "lane keeping, distance pacing, and more." These are not future considerations. These are current consumer and business expectations. They indicate clear and present needs to have an AI strategy and to have ML and training data competency in your company.

The Foundation of Successful AI

Machine learning is about learning from data. Historically, this meant creating datasets in the form of logs, or similar tabular data such as "Anthony viewed a video."

These systems continue to have significant value. However, they have some limits. They won't help us do things modern training data–powered AI can do, like build systems to understand a CT scan or other medical imaging, understand football tactics, or in the future, operate a vehicle.

The idea behind this new type of AI is a human expressly saying, "Here's an example of what a player passing a ball looks like," "Here's what a tumor looks like," or "This section of the apple is rotten."

This form of expression is similar to how in a classroom a teacher explains concepts to students: by words and examples. Teachers help fill the gap between the textbooks, and students build a multidimensional understanding over time. In training data, the annotator acts as the teacher, filling the gap between the schema and the raw data.

Dataset (Definition)

A dataset is like a folder. It usually has the special meaning that there is both "raw" data (such as images) and annotations in the same place. For example, a folder might have 100 images plus a text file that lists the annotations. In practice, a dataset is dynamic and often stored in a variety of ways. From the ML perspective, the dataset is the ground truth distribution of the data that the model will consume (e.g., learning by fitting parameters).

Training Data Is Here to Stay

As mentioned earlier, use cases for modern AI/ML data are transitioning from R&D to industry. We are at the very start of a long curve in that business cycle. Naturally, the specifics shift quickly. However, the conceptual ideas around thinking of day-to-day work as annotation, encouraging people to strive more and more for unique work, and oversight of increasingly capable ML programs, are all here to stay.

On the research side, algorithms and ideas on how to use training data both keep improving. For example, the trend is for certain types of models to require less and less data to be effective. The fewer samples a model needs to learn, the more weight is put on creating training data with greater breadth and depth. And on the other side of the coin, many industry use cases often require even greater amounts of data to reach business goals. In that business context, the need for more and more people to be involved in training data puts further pressure on tooling.

In other words, the expansion directions of research and industry put more and more importance on training data over time.

Training Data Controls the ML Program

The question in any system is control. Where is the control? In normal computer code, this is human-written logic in the form of loops, if statements, etc. This logic defines the system.

In classic machine learning, the first steps include defining features of interest and a dataset. Then an algorithm generates a model. While it may appear that the algorithm is in control, the real control is exercised by choosing the features and data, which determine the algorithm's degrees of freedom.

In a deep learning system, the algorithm does its own feature selection. The algorithm attempts to determine (learn) what features are relevant to a given goal. That goal is defined by training data. In fact, training data is the primary *definition* of the goal.

Here's how it works. An internal part of the algorithm, called a loss function, describes a key part of how the algorithm can learn a good representation of this goal. The algorithm uses the loss function to determine how close it is to the goal defined in the training data.

More technically, the loss is the error we want to minimize during model training. For a loss function to have human meaning, there must be some externally defined goal, such as a business objective that makes sense relative to the loss function. That business objective may be defined in part through the training data.

In a sense, this is a "goal within a goal"; the training data's goal is to best relate to the business objective, and the loss function's goal is to relate the model to the training data. So to recap, the loss function's goal is to optimize the loss, but it can only do that by having some precursor reference point, which is defined by the training data. Therefore, to conceptually skip the middleman of the loss function, the training data is the "ground truth" for the correctness of the model's relationship to the human-defined goal. Or to put it simply: the human objective defines the training data, which then defines the model.

New Types of Users

In traditional software development there is a degree of dependency between the end user and the engineer. The end user cannot truly say whether the program is "correct," and neither can the engineer.

It's hard for an end user to say what they want until a prototype of it has been built. Therefore, both the end user and engineer are dependent on each other. This is called a circular dependency. The ability to improve software comes from the interplay between both, to be able to iterate together.

With training data, the humans control the meaning of the system when doing the literal supervision. Data scientists control it when working on schemas, for example when choosing abstractions such as label templates.

For example, if I, as an annotator, were to label a tumor as cancerous, when in fact it's benign, I would be controlling the output of the system in a detrimental way. In this context, it's worth understanding that there is no validation possible to ever 100% eliminate this control. Engineering cannot, both because of the volume of data and because of lack of subject matter expertise, control the data system.

There used to be this assumption that data scientists knew what "correct" was. The theory was that they could define some examples of "correct," and then as long as the human supervisors generally stuck to that guide, they knew what correct was. Examples of all sorts of complications arise immediately: How can an English-speaking data scientist know if a translation to French is correct? How can a data scientist know if a doctor's medical opinion on an X-ray image is correct? The short answer is—they can't. As the role of AI systems grows, subject matter experts increasingly need to exercise control on the system in ways that supersede data science.[4]

Let's consider why this is different from the traditional "garbage in, garbage out" concept. In a traditional program, an engineer can guarantee that the code is "correct" with, e.g., a unit test. This doesn't mean that it gives the output desired by the end user, just that the code does what the engineer feels it's supposed to do. So to reframe this, the promise is "gold in, gold out" as long as the user puts gold in, they will get gold out.

Writing an AI unit test is difficult in the context of training data. In part, this is because the controls available to data science, such as a validation set, are still based on the control (doing annotations) executed by individual AI supervisors.

Further, AI supervisors may be bound by the abstractions engineering defines for them to use. However, if they are able to define the schema themselves, they are more deeply woven into the fabric of the system itself, thus further blurring of the lines between "content" and "system."

This is distinctly different from classic systems. For example, on a social media platform, your content may be the value, but it's still clear what is the literal system (the box you type in, the results you see, etc.) and the content you post (text, pictures, etc.).

Now that we're thinking in terms of form and content, how does control fit back in? Examples of control include:

- Abstractions, like the schema, define one level of control.
- Annotation, literally looking at samples, defines another level of control.

While data science may control the algorithms, the controls of training data often act in an "oversight" capacity, above the algorithm.

Training Data in the Wild

So far, we've covered a lot of concepts and theory, but training data in practice can be a complex and challenging thing to do well.

What Makes Training Data Difficult?

The apparent simplicity of data annotation hides the vast complexity, novel considerations, new concepts and new forms of art involved. It may appear that a human

---

4 There are statistical methods to coordinate experts' opinions, but these are always "additional"; there still has to be an existing opinion.

selects an appropriate label, the data goes through a machine process, and voilà, we have a solution, right? Well, not quite. Here are a few common elements that can prove difficult.

Subject matter experts (SMEs) are working with technical folks in new ways and vice versa. These new social interactions introduce new "people" challenges. Experts have individual experiences, beliefs, inherent bias, and prior experiences. Also, experts from multiple fields may have to work more closely than usual together. Users are operating novel annotation interfaces with few common expectations on what standard design looks like.

Additional challenges include:

- The problem itself may be difficult to articulate, with unclear answers or poorly defined solutions.
- Even if the knowledge is well formed in a person's head, and the person is familiar with the annotation interface, inputting that knowledge accurately can be tedious and time consuming.
- Often there is a voluminous amount of data labeling work with multiple datasets to manage and technical challenges around storing, accessing, and querying the new forms of data.
- Given that this is a new discipline, there is a lack of organizational experience and operational excellence that can only come with time.
- Organizations with a strong classical ML culture may have trouble adapting to this fundamentally different, yet operationally critical, area. This blindspot of thinking they have already understood and implemented ML, when in fact it's a totally different form.
- As it is a new art form, general ideas and concepts are not well known. There is a lack of awareness, access, or familiarity to the right training data tools.
- Schemas may be complex, with thousands of elements, including nested conditional structures. And media formats impose challenges like series, relationships, and 3D navigation.
- Most automation tools introduce new challenges and difficulties.

While the challenges are myriad and at times difficult, we'll tackle each of them in this book to provide a roadmap you and your organization can implement to improve training data.

The Art of Supervising Machines

Up to this point, we've covered some of the basics and a few of the challenges around training data. Let's shift gears away from the science for a moment and focus on the art. The apparent simplicity of annotation hides the vast volume of work involved. Annotation is to training data what typing is to writing. Simply pressing keys on a keyboard doesn't provide value if you don't have the human element informing the action and accurately carrying out the task.

Training data is a new paradigm upon which a growing list of mindsets, theories, research, and standards are emerging. It involves technical representations, people decisions, processes, tooling, system design, and a variety of new concepts specific to it.

One thing that makes training data so special is that it is capturing the user's knowledge, intent, ideas, and concepts without specifying "how" they arrived at them. For example, if I label a "bird," I am not telling the computer what a bird is, the history of birds, etc.—only that it *is* a bird. This idea of conveying a high level of intent is different from most classical programming perspectives. Throughout this book, I will come back to this idea of thinking of training data as a new form of coding.

A New Thing for Data Science

While an ML model may consume a specific training dataset, this book will unpack the myriad of concepts around the abstract concepts of training data. More generally, training data is not data science. They have different goals. Training data produces structured data; data science consumes it. Training data is mapping human knowledge from the real world into the computer. Data science is mapping that data back to the real world. They are the two different sides of the coin.

Similar to how a model is consumed by an application, training data must be consumed by data science to be useful. The fact that it's used in this way should not detract from its differences. Training data still requires mappings of concepts to a form usable by data science. The point is having clearly defined abstractions between them, instead of ad hoc guessing on terms.

It seems more reasonable to think of training data as an art practiced by all the other professions, that is practiced by subject matter experts from all walks of life, than to think of data science as the all-encompassing starting point. Given how many subject matter experts and non-technical people are involved, the rather preposterous alternative would seem to assume that data science towers over all! It's perfectly natural that, to data science, training data will be synonymous with labeled data and a subset of overall concerns; but to many others, training data is its own domain.

While attempting to call anything a new domain or art form is automatically presumptuous, I take solace in that I am simply labeling something people are already doing. In fact, things make much more sense when we treat it as its own art and stop shoehorning it into other existing given categories. I cover this in more detail in Chapter 7.

Because training data as a named domain is new, the language and definitions remain fluid. The following terms are all closely related:

- Training data
- Data labeling
- Human computer supervision
- Annotation
- Data program

Depending on the context, those terms can map to various definitions:

- The overall art of training data
- The act of annotating, such as drawing geometries and answering schema questions
- The definition of what we want to achieve in a machine learning system, the ideal state desired
- The control of the ML system, including correction of existing systems
- A system that relies on human-controlled data

For example, I can refer to annotation as a specific subcomponent of the overall concept of training data. I can also say "to work with training data," to mean the act of annotating. As a novel developing area, people may say *data labeling* and mean just the literal basics of annotation, while others mean the overall concept of training data.

The short story here is it's not worth getting too hung up on any of those terms, and the context it's used in is usually needed to understand the meaning.

ML Program Ecosystem

Training data interacts with a growing ecosystem of adjacent programs and concepts. It is common to send data from a training data program to an ML modeling program, or to install an ML program on a training data platform. Production data, such as predictions, is often sent to a training data program for validation, review, and further control. The linkage between these various programs continues to expand. Later in this book we cover some of the technical specifics of ingesting and streaming data.

Raw data media types

Data comes in many media types. Popular media types include images, videos, text, PDF/document, HTML, audio, time series, 3D/DICOM, geospatial, sensor fusion, and multimodal. While popular media types are often the best supported in practice, in theory any media type can be used. Forms of annotation include attributes (detailed options), geometries, relationships, and more. We'll cover all of this in great detail as the book progresses, but it's important to note that if a media type exists, someone is likely attempting to extract data from it.

Data-Centric Machine Learning

Subject matter experts and data entry folks may end up spending four to eight hours a day, every day, on training data tasks like annotation. It's a time-intensive task, and it may become their primary work. In some cases, 99% of the overall team's time is spent on training data and 1% on the modeling process, for example by using an AutoML-type solution or having a large team of SMEs.[5]

Data-centric AI means focusing on training data as its own important thing—creating new data, new schemas, new raw data capturing techniques, and new annotations by subject matter experts. It means developing programs with training data at the heart and deeply integrating training data into aspects of your program. There was mobile-first, and now there's data-first.

In the data-centric mindset you can:

- Use or add data collection points, such as new sensors, new cameras, new ways to capture documents, etc.
- Add new human knowledge in the form of, for example, new annotations, e.g., from subject matter experts.

The rationales behind a data-centric approach are:

- The majority of the work is in the training data, and the data science aspect is out of our control.
- There are more degrees of freedom with training data and modeling than with algorithm improvements alone.

When I combine this idea of data-centric AI with the idea of seeing the breadth and depth of training data as its own art, I start to see the vast fields of opportunities. What will you build with training data?

---

5 I'm oversimplifying here. In more detail, the key difference is that while a data science AutoML training product and hosting may be complex itself, there are simply fewer people working on it.

Failures

It's common for any system to have a variety of bugs and still generally "work." Data programs are similar. For example, some classes of failures are expected, and others are not. Let's dive in.

Data programs work when their associated sets of assumptions remain true, such as assumptions around the schema and raw data. These assumptions are often most obvious at creation, but can be changed or modified as part of a data maintenance cycle.

To dive into a visual example, imagine a parking lot detection system. The system may have very different views, as shown in Figure 1-4. If we create a training data set based on a top-down view (left) and then attempt to use a car-level view (right) we will likely get an "unexpected" class of failure.

If the left is your training data and the right is your use case, you are in trouble!

Figure 1-4. Comparison of major differences in raw data that would likely lead to an unexpected failure

Why was there a failure? A machine learning system trained only on images from a top-down view, as in the left image, has a hard time running in an environment where the images are from a front view, as shown in the right image. In other words, the system would not understand the concept of a car and parking lot from a front view if it has never seen such an image during training.

While this may seem obvious, a very similar issue caused *a real-world failure (https:// oreil.ly/HyH0r)* in a US Air Force system, leading them to think their system was materially better than it actually was.

How can we prevent failures like this? Well, for this specific one, it's a clear example of why it's important that the data we use to train a system closely matches production data. What about failures that aren't listed specifically in a book?

The first step is being aware of training data best practices. Earlier, talking about human roles, I mentioned how communication with annotators and subject matter experts is important. Annotators need to be able to flag issues, especially those regarding alignment of schemas and raw data. Annotators are uniquely positioned to surface issues outside the scope of specified instructions and schemas, e.g., when that "common sense" that something isn't right kicks in.

Admins need to be aware of the concept of creating a novel, well-named schema. The raw data should always be relevant to the schema, and maintenance of the data is a requirement.

Failure modes are surfaced during development through discussions around schema, expected data usage, and discussions with annotators.

Failure Example in a Deployed System

Given how new some of these systems are, it's likely that we have barely seen the smallest of the failure cases of training data.

In April 2020, Google deployed a medical AI to help with COVID-19.[6] They trained it with higher-quality scans than what was available during production. So, when people went to actually use it, they had to often retake the scans to try to meet that expected quality level. And even with this extra burden of retaking them, the system still rejected about 25%. That would be like an email service that made you resend every second email and completely refused to deliver every fourth email.

Of course, there are nuances to that story, but conceptually it shows how important it is to align the development and production data. What the system trains on needs to resemble what will actually be used in the field. In other words, don't use "lab"-level scans for the development set and then expect a smartphone camera to work well in production. If production will be using a smartphone camera, then the training data needs to come from one, too.

History of Development Affects Training Data Too

When we think of classic software programs, their historical development biases them toward certain states of operation. An application designed for a smartphone has a certain context, and may be better or worse than a desktop application at certain things. A spreadsheet app may be better suited for desktop use; a money-sending system disallows random edits. Once a program like that has been written, it becomes

---

6 Read Will Douglas Heaven's article, "Google's Medical AI Was Super Accurate in a Lab. Real Life Was a Different Story" (*https://oreil.ly/2Phl2*), *MIT Technology Review*, April 27, 2020.

hard to change core aspects, or "unbias it." The money-sending app has many assumptions built around an end user not being able to "undo" a transaction.

The history of a given model's development, accidental or intentional, also affects training data. Imagine a crop inspection application mostly designed around diseases that affect potato crops. There were assumptions made regarding everything from the raw data format (e.g., that the media is captured at certain heights), to the types of diseases, to the volume of samples. It's unlikely it will work well for other types of crops. The original schema may make assumptions that become obsolete over time. The system's history will affect the ability to change the system going forward.

What Training Data Is Not

Training data is not an ML algorithm. It is not tied to a specific machine learning approach.

Rather, it's the definition of what we want to achieve. The fundamental challenge is effectively identifying and mapping the desired human meaning into a machine-readable form.

The effectiveness of training data depends primarily on how well it relates to the human-defined meaning assigned to it and how reasonably it represents real model usage. Practically, choices around training data have a huge impact on the ability to train a model effectively.

Generative AI

Generative AI (GenAI) concepts, like generative pre-trained transformers (GPTs) and large language models (LLMs), became very popular in early 2023. Here, I will briefly touch on how these concepts relate to training data.

At the time of writing, this area is moving very rapidly. Major commercial players are being extremely restrictive in what they share publicly, so there's a lot of speculation and hype but little consensus. Therefore, most likely, some of this Generative AI section will be out of date by the time you are reading it.

We can begin with the concept of unsupervised learning. The broadly stated goal of unsupervised learning in the GenAI context is to work without newly defined human-made labels. However, LLMs' "pre-training" is based off of human source material. So you still need data, and usually human-generated data, to get something that's meaningful to humans. The difference is when "pre-training" a generative AI, the data doesn't initially need labels to create an output, leading to GenAI being affectionately referred to as the unsupervised "monster." This "monster," as shown in Figure 1-5 must still be tamed with human supervision.

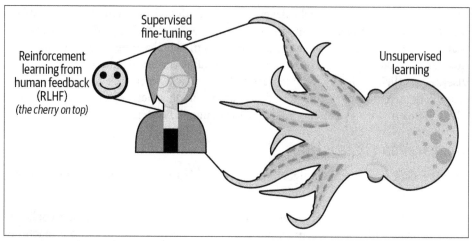

Figure 1-5. Relationship of unsupervised learning to supervised fine-tuning and human alignment

Broadly speaking, these are the major ways that GenAI interacts with human supervision:

Human alignment
Human supervision is crucial to build and improve GenAI models.

Efficiency improvements
GenAI models can be used to improve tedious supervision tasks (like image segmentation).

Working in tandem with supervised AI
GenAI models can be used to interpret, combine, interface with, and use supervised outputs.

General awareness of AI
AI is being mentioned daily in major news outlets and on earning calls by companies. General excitement around AI has increased dramatically.

I'll expand on the human alignment concept in the next subsection.

You can also use GenAI to help improve supervised training data efficiency. Some "low-hanging" fruit, in terms of generic object segmentation, generic classification of broadly accepted categories, etc., is all possible (with some caveats) through current GenAI systems. I cover this more in Chapter 8 when I discuss automation.

Working in tandem with supervised AI is mostly out of scope of this book, beyond briefly stating that there is surprisingly little overlap. GenAI and supervised systems are both important building blocks.

Advances in GenAI have made AI front-page news again. As a result, organizations are rethinking their AI objectives and putting more energy into AI initiatives in general, not just GenAI. To ship a GenAI system, human alignment (in other words, training data) is needed. To ship a complete AI system, often GenAI + supervised AI is needed. Learning the skills in this book for working with training data will help you with both goals.

Human Alignment Is Human Supervision

Human supervision, the focus of this book, is often referred to as human alignment in the generative AI context. The vast majority of concepts discussed in this book also apply to human alignment, with some case-specific modifications.

The goal is less for the model to directly learn to repeat an exact representation, but rather to "direct" the unsupervised results. While exactly which human alignment "direction" methods are best is a subject of hot debate, specific examples of current popular approaches to human alignment include:

- *Direct supervision*, such as question and answer pairs, ranking outputs (e.g., personal preference, best to worst), and flagging specifically iterated concerns such as "not safe for work." This approach was key to GPT-4's fame.
- *Indirect supervision*, such as end users voting up/down, providing freeform feedback, etc. Usually, this input must go through some additional process before being presented to the model.
- *Defining a "constitutional" set of instructions* that lay out specific human supervision (human alignment) principles for the GenAI system to follow.
- *Prompt engineering*, meaning defining "code-like" prompts, or coding in natural language.
- *Integration* with other systems to check the validity of results.

There is little consensus on the best approaches, or how to measure results. I would like to point out that many of these approaches have been focused on text, limited multimodal (but still text) output, and media generation. While this may seem extensive, it's a relatively limited subsection of the more general concept of humans attaching repeatable meaning to arbitrary real-world concepts.

In addition to the lack of consensus, there is also conflicting research in this space. For example, two common ends of the spectrum are that some claim to observe emergent behavior, and others assert that the benchmarks were cherry-picked and that it's a false result (e.g., that the test set is conflated with the training data). While it seems clear that human supervision has something to do with it, exactly what level, and how much, and what technique is an open question in the GenAI case. In fact, some results show that small human-aligned models can work as well or better than large models.

While you may notice some differences in terminology, many of the principles in this book apply as well to GenAI alignment as to training data. Specifically, all forms of direct supervision are training data supervision. A few notes before wrapping up the GenAI topic: I don't specifically cover prompt engineering in this book, nor other GenAI-specific concepts. However, if you are looking to build a GenAI system, you will still need data, and high-quality supervision will remain a critical part of GenAI systems for the foreseeable future.

Summary

This chapter has introduced high-level ideas around training data for machine learning. Let's recap why training data is important:

- Consumers and businesses show increasing expectations around having ML built in, both for existing and new systems, increasing the importance of training data.

- It serves as the foundation of developing and maintaining modern ML programs.

- Training data is an art and a new paradigm. It's a set of ideas around new, data-driven programs, and is controlled by humans. It's separate from classic ML, and comprises new philosophies, concepts, and implementations.

- It forms the foundation of new AI/ML products, maintaining revenue from existing lines of business, by replacing or improving costs through AI/ML upgrades, and is a fertile ground for R&D.

- As a technologist or as a subject matter expert, it's now an important skill set to have.

The art of training data is distinct from data science. Its focus is on the control of the system, with the goal that the system itself will learn. Training data is not an algorithm or a single dataset. It's a paradigm that spans professional roles, from subject matter experts, to data scientists, to engineers and more. It's a way to think about systems that opens up new use cases and opportunities.

Before reading on, I encourage you to review these key high-level concepts from this chapter:

- Key areas of concern include schemas, raw data, quality, integrations, and the human role.
- Classic training data is about discovery, while modern training data is a creative art; the means to "copy" knowledge.
- Deep learning algorithms generate models based on training data. Training data defines the goal, and the algorithm defines how to work toward this goal.
- Training data that is validated only "in a lab" will likely fail in the field. This can be avoided by primarily using field data as the starting point, by aligning the system design, and by expecting to rapidly update models.
- Training data is like code.

In the next chapter we will cover getting set up with your training data system and we'll learn about tools.

CHAPTER 2

Getting Up and Running

Introduction

There are many tools available that help us when we work with data: we have databases to smoothly store data and web servers to smoothly serve data. And now, training data tools to smoothly work with training data.

In addition to tools, there are established processes and expectations for how databases integrate with the rest of your application. But what about training data? How do you get up and running with training data? Throughout this chapter, I'll cover key considerations, including installation, annotation setup, embedding, end user, workflow, and more.

It's important to note why I referenced smoothly working with training data earlier. I say "smoothly" because I don't have to use a database. I could write my data to a file and read from that. Why do I need a database, like Postgres, to build my system? Well, because Postgres brings a vast variety of features, such as guarantees that my data won't easily get corrupted, that data is recoverable, and that data can be queried efficiently. Training data tools have evolved in a similar way.

In this chapter I will cover:

- How to get up and running
- The scope of training data tools
- The benefits you get from using training data tools
- Trade-offs
- The history that got us to where we are today

Most of this is focused on things that will be relevant to you today. I also include some brief sections on history to demonstrate why these tools matter. Additionally, I will also answer other common questions:

- What are the key conceptual areas of training data tools?
- Where do training data tools fit in your stack?

Before we dive in, there are two important themes to mention that you'll see come into play throughout this chapter.

I will often use *tool* as the word even when it may be a larger system or platform. By tool, I mean any technology that helps you accomplish your training data goals. Tool usage is part of the day-to-day work of training data. Throughout the book, I ground abstract concepts into specific examples with tooling. By jumping between high-level concepts and specific implementation samples, you will get a more complete picture.

Practice makes permanent. Like any art, you must master the tool of the trade. With training data, there are a variety of tooling options to become familiar with and understand. I'll talk about some of the trade-offs, such as regarding closed or open source and deployment options, and we'll explore popular tools.

Getting Up and Running

The following section is a minimal viable roadmap to get your training data systems up and running. It is divided, for convenience, into sections. Usually, these tasks can be given to different people, and many can be done in parallel. Depending on a number of factors, it may take many months to get fully set up, and you should consider this in your planning.

If you are starting from a fresh slate, then all of these steps will be applicable. If your team is already progressing well, then this provides a checklist to see whether your existing processes are comprehensive.

Broadly, the overall getting-started tasks include:

- Installation
- Tasks setup
- Annotator user setup
- Data ingestion setup
- Data catalog setup
- Workflow setup
- Initial usage
- Optimization

These steps will be covered in an informative and basic way. Later, "Trade-Offs" on page 41 will discuss other practical considerations like costs, installation options, scaling, scope, and security.

If that seems like it's a lot, well, that's the reality of what's needed to set up a successful system.

In most of these steps, there is some level of crossover. For example, nearly any of the steps can be done through UI/SDK/API. Where appropriate, I'll call attention to common preferences.

Installation

Your training data installation and configuration is done by a technical person, or team of people.

High-level concerns of installation include:

- Provisioning hardware (cloud or otherwise)
- Doing the initial installation
- Configuring initial security items, like Identity providers
- Choosing storage options
- Capacity planning
- Maintenance dry runs, like doing updates
- Provisioning initial super users

Most teams shipping complex, revenue-impacting products do their own installation. This is just a reality of the level of importance of the data, and of its deep connection to end users. Generally, data setup is of a more fluid nature than the installation of the training data platform itself, so data setup is treated as its own section, and we'll cover that next. For now, we'll start by installing Diffgram, commercial open source and fully featured software that can be downloaded from the Diffgram site (*http://diffgram.com*). While you don't need to install it to complete the following examples, you may find it helpful to follow along and get a feel for these exercises in practice. The install welcome screen is shown in Figure 2-1.

The installation process for all tools is always changing, so check the docs (*https://oreil.ly/6WX8Z*) for the latest installation options and guidance. Diffgram, for example, has multiple installation options for "Baremetal," production, and Docker, etc.

At the time of writing, the Docker dev installation prompts are as follows (Shell in an interpreter):

```
git clone https://github.com/diffgram/diffgram.git
cd diffgram
pip install -r requirements.txt
python install.py
```

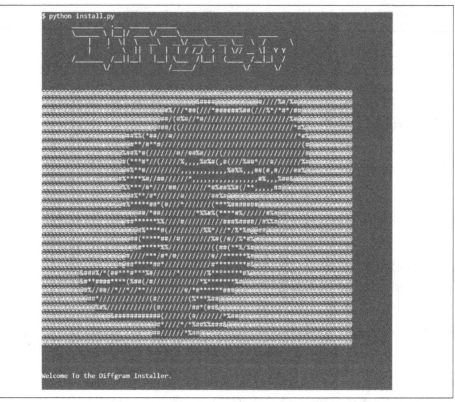

Figure 2-1. Example Diffgram dev installation ASCII art shows you are on the right path

Tasks Setup

Tasks setup is usually done by an admin in conjunction with data science and data engineering personnel.

Steps taken during this stage include:

- Initial schema setup (data science and/or admins)
- Initial human tasks setups (data engineering and/or admins)

Planning takes place around structuring the schema in relevant ways. As these tools grow, there are more and more options, and it is becoming like database design. Depending on the complexity, the schema may be loaded through an SDK or API.

Annotation is covered in more detail in Chapters 3 and 5.

Annotator Setup

Your users (annotators) are often the best people to provide supervision (annotation). After all, they already have the context of what they want. Your users can provide this annotation supervision, and will help to make scale adjustments better than if you were trying to hire ever-larger central annotation teams. These annotators need to be able to have UI/UXs to input their annotations. The most common way to achieve this is through "stand-alone" portal tools.

Portal (default)

Using a stand-alone portal is often easiest. This means that annotators go directly to your installation, e.g., at a web address such as *"your_tool.your_domain.com"* or *"diffgram.your_domain.com,"* to do the annotation. You can then optionally "deep link" to the annotation portal in your application, for example linking to specific tasks based on context in your application. Portals usually offer some form of customization of the look and feel of annotation UI/UXs in addition to functional changes based on the schema. With OAuth plus this customization, you can maintain a seamless look and feel.

Embedded

Your engineering team will need to add code to your application to embed the annotator setup directly inside your application. This is more work but can offer more flexibility and capabilities. Naturally, the capabilities of the development team adding the data setup must be considered; it is far easier to direct users to a known portal, or do API integrations to link the portal, versus deeply integrating and embedding the setup within your application.

Data Setup

You must load your raw data in a useful way to the system:

- Using ingestion tools
- Integrating within your custom application
- SDK/API usage

Workflow Setup

Your data must be able to connect with your ML programs. Common ways to do this include manual processes, integrations (e.g., APIs), or a named training data workflow. A workflow combines multiple action steps to create surfaced processes specific to training data. Integrations with other applications and processes are common ways to connect your data. Some of these steps may be automated, e.g., through pipelines. I cover workflow in more detail in Chapter 5.

Data Catalog Setup

In some way, you must be able to view the results of annotation work, and you will need to be able to access the data at the set level:

- Domain-specific query languages are becoming more popular for training data; or being able to understand raw SQL structure well enough to query directly will be helpful.
- Using training-data-specific libraries for specific goals like data discovery, filtering, etc.
- Even if there are workflows set up, the need for data catalog–type steps remains.

Initial Usage

There is always a period of initial usage. This is where annotators, your end users, data scientists, ML programs, etc. all operate on and with the data:

- Training of users, especially if deployed toward in-house users
- User feedback, especially if deployed toward end users

Optimization

Once the basics are set up, there are many concepts that can be further optimized:

- Day-to-day work of optimizing the schema, raw data, and training data itself
- Annotation proficiency, ergonomics, etc.
- Literal annotation proficiency
- Loading data into machine learning tools, libraries, and concepts
- Contributing to open source training data projects
- General knowledge of available new tools

Tools Overview

Training data tools are required to ship your AI/ML program. Key areas include annotation, cataloging, and workflow:

Annotation

An end user annotates data using annotation tools. The spectrum goes from the annotation being part of your application to stand-alone pure annotation tools. In the most abstract terms, this is the "data entry" side:

- Literal annotation UIs for images, video, audio, text, etc.
- Manage tasks, quality assurance, and more.

Cataloging

Cataloging is the search, storage, exploration, curation, and usage of sets of data. This is the "data out" side, usually with some level of human involvement, such as looking at sets of data. Examples of cataloging activities weave through the whole model training process:

- Ingest raw data, prediction data, metadata, and more.
- Explore everything from filtering uninteresting data to visually viewing it.
- Debug existing data, such as predictions and human annotations.
- Manage the data lifecycle, including retention and deletion, as well as Personally Identifiable Information and access controls.

Workflow

Workflow is the processes and data flow of your training data. It's the glue between annotation, cataloging, and other systems. Think integrations, installations, plugins, etc. This is both data in and data out, but usually at more of a "system" level:

- Ecosystem of technologies.
- Annotation automation: anything that improves annotation performance, such as pre-labeling or active learning. See Chapter 6 for more depth.
- Collaboration across teams between machine learning, product, ops, managers, and more.
- Stream to training: getting the data to your models.

Some products cover most of these areas in one platform.

Training Data for Machine Learning

Usually, machine learning modeling and training data tools are different systems.

The more ML program support that is natively included, the less flexible and powerful the overall system is.

As an analogy, in an MS Word doc, I can create a table, within the constraints of the document editing application. This is distinctly different from the power of formulas that a spreadsheet application brings, which has more freedom and flexibility for calculations but is not a document editor.

Usually, the best way around this is with great integrations, so that systems that focus on training data quality, ML modeling, etc. can be their own systems, but still tightly integrated with training data.

This chapter will focus on the major subareas of training data, specifically assuming that the model training is handled by a different system.

Growing Selection of Tools

There are an increasing number of notable platforms and tools becoming available. Some aim to provide broad coverage, while others cover deep and specific use cases in each of these areas. There are tens of notable tools that fall into each of the major categories.

As the demand for these commercial tools continues to grow, I expect that there will be both a stream of new tools entering the market and a consolidation in some of the more mature areas. Annotation is one of the more mature areas; data exploration, in this context, is relatively new.

I encourage you to continuously explore the options available that may net different and improved results for your team and product in the future.

People, Process, and Data

Employee time in any form is often the greatest cost center.

Well-deployed tooling brings many unique efficiency improvements, many of which can stack to create many-orders-of-magnitude improvements in performance. To continue the database analogy, think of it as the difference between sequential scans and indexes. One may never complete, while the other is fast! Training data tools upgrade you to the world of indexes. Training data tools allow you to perform some key tasks:

- Embed supervision directly in your application.

- Empower people, processes, and data in the context of human computer supervision.

- Standardize your training data efforts around a common core.

- Surface training data issues.

Training data tools similarly provide many benefits that go far beyond handling the minutiae. For example, data scientists can query the data trained by annotators without having to download huge sets and manually filter locally. However, for that to work, a training data system must be set up first.

Your training data learning is an ongoing journey rather than a destination. I bring this up to a level set that no matter how familiar you are or how much time you spend, there is always more to learn about training data.

Embedded Supervision

When we get a spam email, we mark it as spam, teaching the system. We can add words to our spellchecker. These are simple examples of end-user engagement. Over time, more and more of the supervision may be pushed as near the end user as possible, and embedded in systems. If those users are consciously aware that they are annotating, then the quality may be similar to an independent annotation portal. If the users are unaware, there are quality and other risks, as with any end-user data collection.

Human Computer Supervision

As annotation becomes more mainstream, it may be useful to think about how it relates to classic concepts like human–computer interaction (HCI). HCI is how a user relates to and engages with a computer program. With training data I'd like to introduce a concept called human computer supervision (HCS). The idea behind HCS is that you are supervising the "computer." The "computer" could be a machine learning model, or a larger system. The supervision happens on multiple levels, from annotation to approving datasets.

To contrast these, in HCI the user is primarily the "consumer," whereas in HCS, the user is more the "producer." The user is producing supervision, in addition to interaction, which is consumed by the ML program.

The key contrast here is that usually, with computer interaction, it's deterministic. If I update something, I expect it to be updated, whereas with computer supervision, like with human supervision, it's non-deterministic. There's a degree of randomness. As a supervisor, I can supply corrections, but for each new instance, the computer still makes its own predictions. There is also a time element—usually an HCI is an "in the moment" thing, whereas HCS operates at longer time scales where supervision affects a more nebulous, unseen system of machine learning models.

Separation of End Concerns

Training data tools separate end-user data capture concerns from other concepts. For example, you might want to be able to add in end-user data capture of supervision, without worrying about the data flow.

Standards

Training data tools are a means to effectively ship your machine learning product. As a means to this complex end, training data tools come with some of the most diverse opinions and assumptions of any area of modern software.

Tools help bring some standardization and clarity to the noise. They also help bring teams that may otherwise have no benchmark of comparison rapidly into the light:

- Why not have end users provide some of the supervision?
- Why manually version control when you can do it automatically?
- Why manually export files, when you can stream the data you need?
- Why have different teams storing the same data with slightly different tags, when you can use a single unified datastore?
- Why manually assign work when it can be done automatically?

Many Personas

There are many different personas involved with each group having their own common preferences. Common groups and preferences include:

- Annotators expect to be able to annotate as well as the best drawing tools.
- Engineers expect to be able to customize it.
- Managers expect modern task management like a dedicated task management system.
- Data engineers expect to be able to ingest and process huge amounts of data, effectively making it an "Extract Transform Load" tool in its own right.
- Data scientists expect to be able to conduct analysis on it like a business intelligence tool.

Given these personas, training data tooling involves features that resemble portions of: Photoshop + Word + Premiere Pro + Task Management + Big Data Feature Store + Data Science Tools.

Not having the right training data tools would be like trying to build a car without a factory. Achieving a fully set up training data system can only be accomplished through the use of tools.

We have a tendency to take for granted the familiar. It's just a car, or just a train, or just a plane. All engineering marvels in their own right! We similarly discount what we don't understand. "Sales can't be that hard," says the engineer. "If I were President," etc.

By keeping in mind the different personas listed previously you can start to see the breadth of the industrial system needed (and how those personas are served by a system supporting the literal work of annotation and the task management). A suite of annotation interfaces comparable to the Adobe Suite or MS Office. The systems' many distinct functions must work in concert, with input and output directly used in other systems, plus integrated data science tools.

A Paradigm to Deliver Machine Learning Software

The same way the DevOps mindset gives you a paradigm to deliver software, the training data mindset gives you a paradigm to ship machine learning software. To put it very plainly, these tools provide:

- Basic functions to work with training data, like annotation and datasets. Things that it would be otherwise impractical to do without tooling.
- Provide guardrails to level set your project. Are you actually following a training data mindset?
- The means to achieve training data goals like managing costs, tight iterative time-to-data loops, and more.

Trade-Offs

If you are comparing multiple tools for production usage, there are some trade-offs to consider.

In the rest of the book, we continue to cover abstract concepts. In this section I pause to cover real-world industry trade-offs, especially around tooling.

Costs

Common costs include embedded integration, end user input development and customization, commercial software licenses, hardware, and support.

Besides commercial costs, all tools have hardware costs. This is mostly to host and store data, and some tools charge for use of special tools, computer use, and more.

Common cost reductions:

- End-user input (embedded) is materially less expensive than hiring more annotators.
- Push automations onto the frontend as much as possible. This reduces the cost of server-side computation, with the trade-off of running the model locally, which may also cause local lag time.
- Separate true data science training costs from annotation automation.

Common licensing models include unlimited, by user, by cluster, or other more specific metrics.

Some commercial open source products may allow a trial to build out a case for a paid license, and some are free for personal or education use.

Most SaaS (software as a service) training data services have severe limits in the free tier. And some SaaS services may even have privacy clauses that permit them to use your data to build "mega" models that are to their benefit.

Installed Versus Software as a Service

The volume of training data *data* is very high as compared to other types of software, tens to thousands of times more than many other typical use cases. Second, the data is often of a sensitive nature, such as medical data, IDs, bank documents, etc. Third, because training data is like software code, and often contains unique IP information and subject matter expertise, it's very important to protect it. So to recap, training data is:

- High volume
- Sensitive
- Contains unique IP information

The result of this is that there is a massive difference between using a tool you can install on your own hardware, and using SaaS. Given these variables, there is a clear edge to training data products that can be installed on your hardware from day one. Keep in mind, "your hardware" may mean your cluster in a popular cloud provider. As packaging options improve, it becomes easier and easier to get up and running on your own hardware.

This is also another area where open source really shines. While SaaS providers sometimes have more expensive versions that can be locally deployed, the ability to inspect the source code is often still limited (or even zero!). Further, these use cases are often rather rigid: it's a preset setup and requirements set. Software that's designed to run anywhere can be more flexible to your specific deployment needs.

Development System

There is a classic debate of "build or buy." I think this should really be "customize, customize, or customize?" because there is no valid reason to start from scratch at this point with so many great options already available as starting points. Some options, like Diffgram, offer full development systems, that allow you to build your own IP on top of the baseline platform, including embedding annotation collection in your application. Some options have increasing degrees of out-of-the-box customization. Open source options can be built on and extended.

For example, maybe your data requirements mean the ingestion capabilities or database of a certain tool isn't enough, or maybe you have a unique UI requirement. The real question is, should we do this ourselves, or get a vendor to do it for us?

Sequentially dependent discoveries

A mental picture I like to think about is standing at the base of a large hill or mountain. From the base, I can't see the next hill. And even from the top of that hill, my vantage is obscured by the next, such that I can't see the third mountain until I traverse the second, as shown in Figure 2-2. Essentially, subsequent discoveries are dependent on earlier ones.

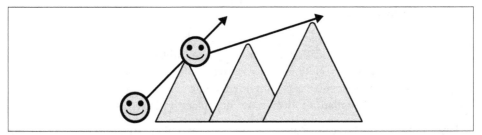

Figure 2-2. Sight lines over hills: I can only see the next mountain

Training data tools help you smoothly traverse these mountains as you encounter them, and in some cases they can even help you "see around corners" and give you a bird's-eye view of the terrain. For example, I only understood the need for querying data, e.g., to stream a slice of data instead of file-level export, once I realized that approaches to annotate data often don't align with the needs of a data scientist, especially on larger teams over large periods of time. This means that no matter how good the initial dataset organization is, there is still a need to go back afterwards and explore it.

Training data tools are likely to surprise you with unexpected opportunities to improve your process and ship a better product. They provide a baseline process. They help avoid thinking you have reinvented the wheel, only to realize an off-the-shelf system already does that and with a bit more finesse. They improve your business key performance indicators by helping you ship faster, with less risk, and with more efficiency.

Now of course, it's not that these tools are a cure-all. Nor are they bug free. Like all software, they have their hiccups. In many ways, these are the early years of these tools.

Scale

Disney World operates differently from a local arcade. Likewise, when it comes to data projects, what works for Disney won't work for the arcade, and vice versa.

As I cover in Chapter 8, at the extreme end of the scale, a fully set up training data system allows you to retrain your models practically on demand. Improving the speed of time-to-data (the time between when data arrives and when a model is deployed) to approach zero can mean the difference between tactical relevancy or worthlessness.

Often the terms we are used to thinking about for scale with regular software aren't as well defined with supervised training data. Here I take a moment to set some expectations with scale.

Why is it useful to define scale?

Well, first, to understand what stage you are in to help inform your research directions. Second, to understand that various tools are built for different levels of scale. A rough analogy may be SQLite versus PostgreSQL. Two different purposes from the simple to the complex (a small self-contained single-file database versus an advanced powerful dedicated installation). Both have their place and use cases.

A large project will typically have the greatest percentage of annotation from the embedded collection. A superlarge project may go through billions of annotations per month. At that point, it's less about training a singular model, and more about routing and customizing a set of models, or even user-specific models.

On the other side of the coin, for a small project, a data discovery tool may not be relevant if you plan to use 100% of your data anyway. However, it's almost always good practice to perform at least basic data discovery and analysis. An example of this is in Chapter 1, where we discuss a parking lot detection system trying to work with images taken from completely different perspectives (which also means data distributions). That example showed the importance of understanding the raw data and alignment of training data and production data. More technically, the distribution (how data is clustered around specific values or distributed in a higher dimensional space) of raw data training data should be aligned to the production distribution.

For mid-scale and up, you may prefer that your team goes through a few hours of training to learn the best practices of more complex tools if they are going to work with them day in and day out.

So what makes scale so hard? First, most datasets in the wild don't really reflect what's relevant for commercial projects, or can be misleading. For example, they may have been collected at a level of expense that would be impractical for a regular commercial dataset. This is part of why a "dataset search" doesn't really make that much sense as a starting point. Most public datasets aren't going to be relevant to your use cases, and, of course, will not be styled to each of your end users. Further, most companies keep the deep technical details about their AI projects pretty hush-hush, to a degree noticeably different from more common projects.

All projects have very real challenges, regardless of scale. Table 2-1 segments projects into three buckets, and speaks to common properties of each.

Table 2-1. Data scale comparison

Item	Small	Medium	Large
Volume of annotations at rest in a given period of time.	Thousands	Millions	Billions
Media complexity	Usually singular, e.g., images	May be multimodal, e.g., pairs of images (driver's license front and back)	High media complexity, e.g., complex media types or multimodal
Annotation Integration, e.g., linking tasks to the training data system	Not required	Depends	Likely required
People annotating (subject matter experts)	A single person or small team	End user and a medium-sized team	End users primarily, and multiple teams of people
People with data engineering, data science, etc., hats	A single person	A team of people wearing various "data hats"	Multiple teams of people
Modeling concepts	A singular model	A set of named models	"Routing" or channeling data automatically to various models; per-user customization and modeling
Schema complexity	A few labels or a few attributes; <100 elements	Sets of attributes, may be thousands of elements	As complex as product needs dictate
Revenue impacted	No $ amount formally attached or pure research	Millions or tens of millions of dollars impacted by work	Hundreds of millions or billions of dollars impacted by work
System load QPS (queries per second)	<1 QPS	<1000 QPS	>1000s of QPS

Item	Small	Medium	Large
Chief concerns	Ease of getting started and ease of use; cost of tooling (may be no or low budget for tooling)	Effectiveness of tooling; support and uptime of tools; starting to think about optimization; may be planning a transition to major scale	Embedded data collection; volume of data ("scale"); customization; security; inter-team issues; assumption that each team is already doing optimizations they are familiar with

Of course, there are many exceptions and nuances, but if you are trying to scope out a project, this is a good starting point. As you trend toward larger use cases, the following becomes more important:

- Embedded end-user annotation
- Normal system scaling concerns
- Engaging with multiple teams in a standardized way, such as having a similar process for multiple data types

Transitioning from small to medium scale

This also applies to planning a mid-scale system from scratch. Here are a few things to think about:

- Workflow
- Integrations
- Use of more data exploration tooling

As you've seen, scale has many concerns and considerations that you need to account for with your directional planning. Not all of those concerns will apply or be actionable yet, but it's good to be aware of them.

Large-scale thoughts

If you plan to be operating at large scale, there are a few thoughts to keep in mind.

How can we push more of the supervision to the end user? A central team without an embedding plan is a variable cost that increases relative to the scope of the project, quality expectations, and users. However, if the focus of ongoing maintenance is on embedding it with the end user, the central team can be thought of as a fixed startup and quality assurance cost.

What's the velocity at which the data moves through the system? How long does it take to go from new data, to upgraded supervised data, to a new model?

Over the last few years the commercial tooling market has changed dramatically. What was completely unavailable years ago may now be an off-the-shelf option. It's a great time to rethink what unique value-add each team is providing Can you more readily customize an ongoing project than do all the plumbing yourself? Do you really need to build this in-house? Rethinking about aligning with open source standards is especially relevant for large teams that may have formed well prior to those standards being available.

Do we really have to duplicate this data? Is there a more central way we can store this data as it moves around these stages? If you draw out the various parts of your ML and training data, you may be very surprised how many times data is unnecessarily transmitted (e.g., via events) or even duplicated at rest.

Do you have an existing classic (discovery-focused) ML system? Are the concepts and awareness there relevant to this new form of human supervision? Naively speaking, you can think of a given datastore as a base layer, upon which data forks into supervised cases and discovery cases. So, any previous plan you made for your overall ML architecture, if it didn't take supervised annotation into account, needs to be reworked.

Supervised ML is so different that trying to shoehorn it into existing architecture will not work well. Instead, the team needs to create a new ML architecture plan that places supervised learning with its own architecture, with its own process and logical paths. To recap, a great step toward a stronger large-scale architecture is dedicated paths for handling human-supervised data.

How many people does it take to surface issues and make a correction to a model? For example, a worst practice would be something like: Manual feedback process, then a central human annotator, then an ML engineer, then a manager. In that process, it could take months to ship the next iteration of a model. (Imagine if a central team had to be called in every time a user wanted to add a word to their spellcheck dictionary.)

A best practice would be integrated annotation, process-based retraining,[1] automatic validation, and spot checks by a central QA team. This is part of a process of moving from "one-time" human-supervised datasets to continuous human supervision delivery, as far as this is reasonable for your context. Note that continuous updating of human supervision, continuous model training, continuous deployment, etc., may all be different (if related) things, depending on your context.

---

1 The level at which an automatically retrained system should enter production is debatable and context dependent. As with nearly any automation, a larger automation needs greater system-level validation, social understanding, and emergency controls. A more detailed treatment of this is needed for data science and systems-level roles but is well beyond the scope of this book.

What is your policy for the sharing of end-user supervision data? Is there a declared process for when and how shared datasets go into production? For example, this could be dedicated logic in your application, where your end users, super users, etc. have control. Or for system-wide data, it may be similar to approvals for pull requests. You may already have model deployment flows. It's also important to look at the actual data it's being trained on.

Assess the return on investment (ROI) of your data, especially of incremental additions or deletions. At larger scales, it becomes more practical and useful to consider the ROI of new data relative to business goals. For example, you could ask questions like, What percent of our data are we actually using? How much revenue is affected by this dataset?

What does the shape of the data really look like? For example, if you are doing a request/response cycle for every single image, audio file, etc., does that really make sense? Instead, can the data be queried in a central place and then streamed?

Organizations need to ensure that their data governance policies are actually being implemented across teams. Datasets should be stored with the same awareness of expiration controls as the individual elements that comprise them. Managers need to align teams with the tool configurations they'll need to solve problems.

Installation Options

Choosing a package, storage, and database are key parts of installation configuration.

Packaging

Training data tools come in a variety of packages. The way the code is packaged is sometimes an indication of its target audience, such as small, medium, or large projects. I will briefly introduce some common packages:

Single Language-Specific Package
> Some tools may install a single package, for example a Python package. Usually, these are single-point "plugins" and not complete systems. These tools usually only work for small projects.

Docker
> Many tools will require a Docker container, multiple containers, or similar. Docker is a way to package software. Docker Compose is a way to group multiple packages. In theory, any time the Docker images are provided, you can manage those images as you see fit.

Kubernetes (K8s)

K8s orchestrates containers. This is the default recommendation for production, although there are many other options. The major cloud providers have noticeably different implementations of Kubernetes. Specifically, what may take hours of work on one platform is sometimes much easier on others. Training data often represents an unparalleled volume of data, so expectations around data access, storage, and usage are new, and often don't align with many pre-optimized cloud examples.

Storage

You will need to consider where in the world the system is going to be deployed. If you have users in another country, how does that impact your performance and security goals? If a cloud storage option is not available, what types of local options will meet your needs?

Database

Diffgram uses PostgreSQL by default. Many other databases are available. Usually there are at least three different groups of people involved in setting up and using the system:

- Admins
- Technical (engineering, data science)
- Annotators

Data configuration

There are various configurations to be aware of for specific media types. For example, for videos, you'll need to decide whether to store individual frames on demand or not. More generally, there may be choices about what "artifacts" to be stored, such as thumbnails or web-optimized versions:

Versioning resolution

How many versions—and this could potentially be all—of previous annotations are required? Should every change be recorded? In some systems, this may be critical, or simply a useful feature. As a rule of thumb, turning on complete versioning will likely result in at least 80% of the database being composed of these soft-deleted annotations.

Data lifecycle

Do you have to delete the data within a certain period of time? Or must you retain it for a certain length of time? Can some be automatically archived after a period of time? Your team will have to make these determinations.

Annotation Interfaces

Naturally, there needs to be an interface that a human uses to instruct and supervise the machine. Both portal-based and embedded annotation interfaces continue to evolve. In some ways, interfaces are trending toward having relatively similar feature sets, but this remains a very opinionated area. In my opinion, one of the most important things is how well the interface is embedded and presented to the end user. That way, the user can provide meaningful supervision in the most relevant context.

The spectrum of the complexity of types of interfaces and varieties of yet-to-be standard expectations together make discussions around annotation interfaces challenging. For example, annotations centered around attributes (like a regular form) are generally more straightforward than video and 3D systems. However, "forms" themselves can be quite complex. And even within the realm of video, the expectations ranging from being able to merely play a video as a reference point, to annotating specific moments in time, as well as more complex cases, are all different in scope.

Circling back to our project scale concept, for a small-scale project, the overall stylistic feel and out-of-the-box experience may matter. With a large-scale project, the embedded interface will be customized and more likely engineered to specific requirements—for example, choosing which components show up where, color and style (CSS) themes, etc. So, for a large project, the tools' ability to be customized, engineered, and further developed is important.

Modeling Integration

Your training data system needs to communicate with your ML modeling systems. With Diffgram, this is done through API integrations or workflows (covered in later chapters). While sometimes modeling systems may present some superficially similar views, such as outputs with bounding boxes, they usually do not support serious training data work. Modeling integration is related to streaming data, but they are different concepts.

Multi-User versus Single-User Systems

Modern systems, like Diffgram, are multi-user by default. Systems for a single user (e.g., sloth) are usually not part of the modern training data paradigm and are likely focused on the pure user interface portion. The primary reasons for needing multi-user capability are the ease of access to subject matter expertise and ability to handle a larger volume of data. Systems can be operated as a limited prototype by a single user or may run on a single local machine for testing purposes. Much of this chapter is focused on multi-user and team-based systems.

Integrations

There are a few broad categories of integrations:

- System level, one-time setup
- Plugins
- Installable things that run on frameworks or on your training data platform's hardware

Parts of training data tooling are buried in the technical stack, while others are surfaced to end users.

The most basic concept is that you must be able to get raw data and predictions in and annotations out. Considerations include:

- Hardware. Will it run in my environment? Will it work with my storage provider?
- Software infrastructure. Can I use the training systems, analytics, databases, etc. I want with it?
- Applications and services. How well does it integrate with my systems? Backend and frontend?
- Plugins
- What types of custom integrations through APIs and SDKs are available?
- How do I get the data back and forth to the training data system?

Some systems offer greater degrees of integrations. UI-based interactions with integration processes are not just for setting up keys, but also for pulling and pushing the data.

Scope

As this ecosystem continues to evolve, there is a broadening boundary between the scope of users and data the tooling is designed to work with. Some tools may cover multiple scopes. In general, tools lean either toward single users or truly many users.

As shown in Figure 2-3, one way to think of this is as a continuum with two major poles—point and suite solutions.

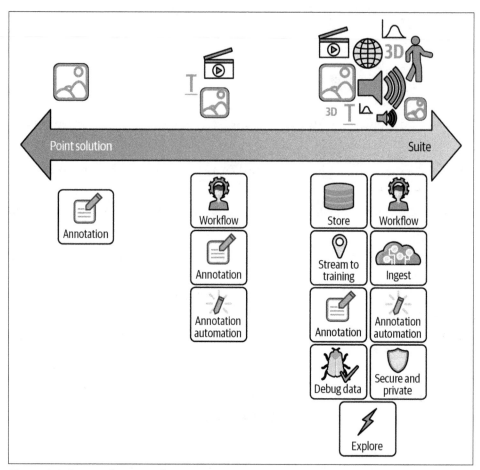

Figure 2-3. Point solution and suite continuum

 Some of these icons are explained throughout the book. Naturally, any system will have some concept of input and output. So when we have an "Ingest" icon, it's there to indicate something that an entire team would work on at a big company. Further icons like "Secure and private" refer to security features like blurring identifying features, personally identifying information (PII), etc., not the general concept of security.

Platform and suite solutions

Platforms and suites are the best options for mid-sized and large teams and companies with multiple teams. Within this category, there are various approaches taken by different service providers. From a broad perspective, understanding the main differences in the "psychology" of these systems can help when it comes to choosing which one to use:

Some view training data as a dedicated discipline
Even if they have other integrated data science products, services, etc., they draw a clear line about what's training data and what's not. Other organizations might not show such a conceptual distinction.

Some offer a suite of media types and lateral supports
Usually, you can tell if it's a broader system because it will cover more—or even all—of the media types. Similarly, for lateral supports like storing, streaming, training, exploring, etc., there will likely be more coverage provided by some solutions. However, keep in mind that even the most advanced and largest platforms have gaps.

Breadth and depth
Further expanding on support for media types, some solutions may offer great coverage on media types, but only to a relatively superficial depth. As a solution leans more toward this end of the spectrum, its depth of offerings in each category continues to grow.

Customization
The big product difference here is that often the developers of these tools assume that they will be customized, either with more built-in customization options through configuration, or with more hooks and endpoints, to naturally customize it through code. Thus, there are quite a few options available.

Decision-making process

Now that you've got the lay of the land, I'll walk you through an example decision-making process. Generally, systems designed for a large scale require consideration of several key factors:

Customization
Virtually everything is up for user configuration, from how the annotation interface looks, to how workflow is structured, etc.

Installation
It's assumed that installations will be done by, or at least overseen by, the customer. Who has the encryption keys, where the data is stored at rest, etc. are part of the discussion. Expect dedicated and clear security discussions.

Hardware
Performance expectations and capacity planning need to be done. Any software, no matter how scalable, still requires more hardware to scale.

Use cases

Many users, teams, data types, etc. should be anticipated.

Data science integrations

Typically overly pre-packaged model training does not support large-scale needs. Instead, focus on integrations to your dedicated data science teams.

Cautions

As with any system, there are some elements that you'll need to be aware of and plan for:

- These systems can be very complex and powerful. They usually take plenty of time to set up, understand, and optimize for your use case. Even obvious "golden path" routes will take time to optimize.
- Direct comparisons to smaller systems may be nonsensical. For example, to meet the multistakeholder goals of a larger system, such as compliance, some things may take longer than what would be obvious in a lighter context.
- Large systems, even with potentially stronger quality controls, have more bugs. What may be hard to break in a small system (due to its simpler nature) may break in a larger system due to the complexity.

Point solutions

On the other side of the continuum from suites are point solutions. Let's walk through what these are, how they can be used, and considerations you should be thinking of:

- Distinguishing features:
  - Often mix training data and data science features; for example, it may be pitched as "End to End" or "Get a Model Trained Faster," which both refer more to data science concepts than training data.
  - Focus on a single or a small handful of media types, e.g., images only.
  - For single users or small teams; this usage assumption cascades to features around who creates labels, ease of use, etc.
  - Software as a service or deployed locally on one machine.
- Usage:
  - For large-scale projects, point solutions can be used for proof of concepts before committing resources to a larger platform. In some cases they may have enough features to be used directly.
  - Usually, by their nature of being simpler, these tools are faster to set up and get a "result." Whether it will be the result you want is often more questionable.

— Having some form of automatic training built in. Automatic training is not an automatic negative; however, usually mid-sized and larger teams want more control, and so may prefer to develop their own training.

— Sometimes point solutions can be of great quality in their specific domain.

- Cautions:

  — These tools usually limit—either by technical or intentional political limits— what type of results can be achieved. For example, they may have a method to train bounding boxes, but not keypoints. Or vice versa. This extends to media types too; they may have a method for images, but none for text.

  — Usually are not appropriate for better-resourced teams. May be lacking many of the major feature areas—such dedicated tasks as workflow functions, ingestion, arbitrary storage and query, etc.

  — Often are not very expandable or customizable relative to heavier-weight solutions.

  — Security and privacy are usually limited. As a specific example, the terms of service may allow these companies to use the data you create to train other models, or sometimes projects are public by default if an organization is not paying, etc. Ultimately, there must be trust in the service provider to handle your data properly.

  — While the quality may be high, the need to string together the point solution with other tools often creates extra work. This is especially prevalent in a larger firm where the tool may be appropriate for one team but not another.

- Cost considerations

  — These types of tools often have a "long tail" of costs. They may have a cost per annotation. Or, it may be free to train the model, but a cost applies to serve it (and no option to download it).

Tools in between

Generally, most tools trend toward one of the ends of the spectrum, either the smaller, as already mentioned, or the larger use cases, as I will cover in a moment. There are also some tools that are somewhat in between either of those poles.

Generally, there are a few key things to look for in a solution:

- More awareness of training data as a separate, standalone concept.

- More awareness of multiple solution paths. Less "one true path" and more flexibility.

- Greater percent of landscape coverage. For example, may have more integrations and flexibility.

- More enterprise-friendly concepts. May offer local installation or customer controlled installations. More focus on customization and function over "golden path" mentalities.

There may be some contractual guarantees around data added. These tools may be able to provide serious results and be appropriate if your team has outgrown smaller tools but doesn't yet have resources for larger tooling.

A suite is not automatically better. However, it is usually hard for smaller tools to "step up" to a higher level, whereas most larger systems can often be used only in part, and they fit this middle path quite well.

The best platforms offer a solution in between the two extremes of "brittle single AutoML" and "do nothing."

Essentially, this means focusing on the human computer supervision side; how to get data to and from machine learning concepts. You'll need to consider how to run your own models, schedule resourcing, and integrate with other systems such as AutoML and dedicated training and hosting systems, etc.

Hidden Assumptions

Training data tools bring many benefits and are of critical importance. To reap the benefits, however, you'll need to consider a few common assumptions. Some of these are usually true, others usually false:

True: The team is important.
> End users, admins, dedicated annotators, engineers, product developers, executives, and more. This is a product that gets touched by many people in the organization, often with very different goals, concerns, and priorities, and it's crucial to incorporate their input.

True: You have to have someone technical on your team.
> Someone needs to install, set up, and maintain the system. Even for 100% service-based tools with the latest wizards, there's still an assumption that at least one person is technically proficient and one person understands training data.

True: You have to do setup.
> Most of this tooling requires setup. While there are a variety of tools that carve out a narrow defined area and reduce the amount of effort required, that's not really data programming, but just consuming a narrowly defined service.

True: You have to consider cost.
> All tools have some form of costs. Even modern commercial open source tools have a licensing cost, and all have hardware and setup costs.

True: You'll need time to get oriented.

The complexity of some of this tooling is quite astounding. As of 2022, open source Diffgram has over 1,400 files and 500,000 lines of code.

False: You must use graphics processing units (GPUs).

Model training often benefits from having a processing accelerator like a GPU. However, actually using a model in automations does not require a GPU. Also, training in the context of a limited dataset does not benefit as much from GPU power because of its smaller size.

False: You must use automations.

Automations are sometimes useful, but they are not required. Improper use can have negative results, with bad feedback loops often produced.

Security

According to a 2022 Linux Foundation report, "Security is the #1 priority that influences what software an organization will use. License compliance is the #2 priority."[2]

Security architecture

For high-security installations, it is usually best to host your own training data tools. This allows you complete control to set your own security practices. You can control the encryption access keys and locations of all aspects of the system, from network to data at rest. And, of course, you can then set your own custom security practices.

Attack surface

Installation is the starting point for considering the attack surface you may be exposing, since networking is 101 for cybersecurity. The attack surface of an inaccessible network is low. So for example, if you already have a hardened cluster, you can install your software and use it within that network. If you don't have a hardened cluster available, you can make a new cluster specific for training data.

Security configuration

Your security posture is dependent on your configuration. For example, you can store objects by reference or ingest them directly into a defined bucket. You can use OIDC (OpenID Connect) or not. You can consider the specific implementation of BLOB signing and configure it accordingly.

---

2 Stephen Hendrick, "Software Bill of Materials (SBOM) and Cybersecurity Readiness" (*https://oreil.ly/7wKiR*), *The Linux Foundation*, January 2022, 6.

Security benefits

There are many security benefits of an installed solution. These include:

- You can set real security, including all keys, based on your real and current security posture.
- You control network security, the annotations database, the raw data, everything.
- You control the entire keychain.
- You are aware of the other threats and you can take action, such as pinning specific versions.
- You can usually inspect the source code.

User access

One of the first things that comes to mind is often users' abilities to access samples. Consider a company that has a smart assistance device. Perhaps a reviewer listens to audio data when the device misfired and the microphone came on accidentally.

Or consider someone correcting a system to detect baby photos, etc. There are many levels to consent, and it needs to be carefully considered.

On the consumer side, there are generally a few general categories regarding consent:

- Do not have consent to use directly (anonymized only)
- Do have consent to use to train models—may be limited by time
- Have consent, but data may contain personally identifiable information that may be sensitive if included in the model

On the commercial side, or with more business-to-business type applications, systems may include confidential customer data. This commercial data may potentially be more valuable than any single consumer record.

In addition, there may be government regulations such as HIPAA or other compliance requirements to consider.

Complicated, right?

Other day-to-day considerations that may come up:

- Can annotators download the data to their local machine?
- Should an annotator be able to access records after completing (submitting) them? Or are they locked out by default?

On the software side there are generally two major models that most approaches fall into:

- Task-only availability. This means that as an annotator user, I can only see my current assigned task (or set of tasks).
- Project level. As an annotator, I can see a set of tasks, or even multiple sets of tasks.

As a project administrator the two big decisions are basically around:

- Structuring the data flow so that only data that is tagged as having consent, and/or meets other PII requirements, ever enters the annotation task flow at all.
- Deciding at what level annotators see tasks.

Data science access

Naturally, data scientists must access this data at some point to do work on it. Often data scientists get a fairly free hand to "look at" the data. A more strict system may allow only sending a query and receiving a sample, with the bulk of the data being sent directly to the training apparatus. This bypasses the data scientist's local machine or user's specific server.

It's worth considering that a single breach of a data scientist's access is often more severe by many orders of magnitude than that of annotators. An annotator, even if you are able to bypass various security mechanisms and store all the data you see, you may only see a small portion of the data of a large project. However, a data scientist may have hundreds of times more access, which the bad actor would gain.

Root-level access

A super-admin-type user, IT administrator, etc., may have some levels of root system access. This may be classified as a super admin in the application, and they may have direct access to databases. This is the most privileged access and should be carefully controlled.

Explainability

Regina Barzialy, professor at MIT's Computer Science and Artificial Intelligence Laboratory (CSAIL), says:

> It's like a dog, which can smell much better than us, explaining how it can smell something. We just don't have that capacity. I think that as the machines become much more advanced, this is the big question. What explanation would convince you if you on your own cannot solve this task?

The concept of explainability (*https://oreil.ly/ipeRW*) is important, but is usually reserved more for the machine learning model analysis side.

Open Source and Closed Source

Open versus closed source is an argument as old as software, but it's one that has become especially important recently. I'd like to take a moment to highlight some specifics I have seen relative to this concept as it pertains to training data.

Open and closed source annotation deserve special consideration in the rapidly changing training data landscape because the majority of this new generation of tooling is closed source.

There have been many open source annotation tooling projects—some over 10 years old. However, in general, most of those projects are no longer maintained, or are very niche and not general-purpose tools. Currently, the two general-purpose "second-generation" annotation tools that are open source are Diffgram and Label Studio. To be sure, there are many other tools—but most are focused on very specific considerations or applications.

Open source software has many advantages—especially in this privacy-focused area. You can see exactly what the source code is doing with your data and be sure there is no nefarious activity afoot.

Open source does have some disadvantages. Most notably, the initial setup of the system itself may be more difficult (not the application setup itself, which is similar in either case, just the actual installation of the overall software).

The commercial costs of both open source and closed source software may be similar; just because the code is open does not mean the license is unlimited. Ease of use is often similar in the context of commercially backed projects.

The costs of hosting open source are controlled by you. In general, the cost of hosting is rolled into the cost paid to a commercial provider. This is a nuanced trade-off, but in practice it is often similar at small and medium scales. At great volumes, usually the more control you have, the better it is for you.

Open source may tend to have greater compatibility, since often there is more use from free users—who still run into issues and write tickets. This can mean less technical risk.

Costs are also similar. Commercially backed open source projects usually require an upgrade to a paid version at some point during commercial use. Sometimes there may be an option to forgo paying, but at the very least that means less support.

Choose an open source tool to get up and running quickly

Some tools install in dev in minutes and a moderate production setup in hours or a few days. Most have optional commercial licenses that can be purchased. This is faster than talking to a sales team and gives a truer account than limited SaaS trials.

See the forest from the trees

Environment setup and adjusting initial expectations is one of the hardest things. It's easy to over-fit on perceived setup/first impressions when much of the value is delivered over a long period of time and to multiple user groups.

Capability over optimizations

What is an optimization for some may be suboptimal for others. For example, an extra "confirm" prompt when completing a task may feel like a huge burden to some, while for others it's a crucial step.

Consider that Excel has over 200 popular shortcuts. My guess is most users only know a small fraction of them, yet are able to use Excel perfectly fine for their jobs. Some people, on the other hand, get very concerned about optimizations, like the specifics of hotkeys.

As annotation continues to become more complex, and new users enter the picture, there is a shift away from shortcuts. The field is moving more toward making sure the UI has the capabilities on offer and that it shows the user a reasonable context for the optimizations so they can make use of those capabilities.

Ease of use in different flows

The ease of updating existing data is often very different from creating all-new annotations.

Vastly different assumptions

I tried one popular annotation UI where the delete key deleted the entire series across all video frames. This would be like painstakingly crafting an entire spreadsheet, only to bump the delete key and have it delete the entire sheet! Even though I was just testing, did I ever get a jolt when that happened!

Of course, someone else could argue that it's easier to use, since I need only select an object and click delete. I don't have to worry about the concept of a series or that it appears in multiple frames. Again—what's right here will depend on your use case. If you have complex per-frame attributes, a single delete for what could be *days* of work is probably bad. Conversely, if you have a simple instance type, in some cases perhaps it's desired.

Again, customization by admins and users comes to the rescue. Do you want to see the previous annotation in the next frame of the video? Or don't you? Choose what's right for you, then set it and forget it.

Look at settings, not first impressions

Even seemingly simple things, like the font size, position, and background of label tags, are all very dependent on the use case. For some, seeing any label may get in the way. For others, the entire meaning is in the attributes, and not showing it slows down progress significantly.

It's the same with polygon size, vertex size, etc. For one user, they may be unhappy if the polygon points are hard to grab and move, while another may want no points at all so that a segmentation line on a medical image can be seen perfectly.

Is it easy to use, or just lacking features?

Another trade-off is that some vendors simply decline to enable features by default, requiring each flow to be planned out. For example, this may mean that instance types aren't available in video, or settings may not exist, etc. When evaluating, think hard about the ongoing usage and needs for more complex scenarios, and whether the service will be able to handle it.

Customization is the name of the game

There is increasingly an expectation of customization capabilities at all levels of software. From embedding interfaces, annotation settings and admin configurations, to actually changing the software itself—we all assume we'll be able to make it our own at every level. Try to be aware of what's "hard" and what's "easy" for your given provider. As a small story, a user noticed that when a task was already completed, pushing a recently added "defer task" button led to a poorly defined state in the system. I agreed this was an issue. The fix was one line of code—a single `if` statement.

For example, for a closed source provider, adding a new storage backend may be a low priority. With an open source project, you may be able to contribute this yourself, or encourage others in the community to do so. Also, you may be able to better scope out and understand the impacts of the changes and costs involved.

On the enterprise side of it, try to understand what the core benefit of the software really is for your use case. Is it a completely integrated platform? Is it the data storage and access layers? Is it the workflow or annotation UI? Because some of these tools vary so dramatically in scope and maturity, it can be hard to compare them. One tool may, for example, have a better spatial annotation UI, but be substantially lacking in many other dimensions, like the ability to update data, ingest, query data, etc.

On the other hand, if a vendor doesn't offer major features like data querying, streaming, wizard-based ingestion, etc., those may all be multimonth projects, multiyear epics, or even never be added at all. Because this is a new area, with vastly different assumptions and expectations, I really encourage you to first consider the major features, and then look at the speed of updates and execution on improvements. A vendor that can adapt quickly is especially valuable in this new area.

Another user had experienced, in a different UI, that deleted singular points were not "recoverable," meaning that if, say, a hand was occluded on a keypoint figure, and they marked it that way, if they went to undo it, they couldn't get it back. In Diffgram, the way the system was set up, it was easy to maintain this feature on a per-point basis.

History

Open Source Standards

By my estimate, in 2017, there were likely less than 100 people in the world working on commercially available tools for training data. In 2022, there are over 1,500 people across at least 40 companies working directly for training data–focused companies. Sadly, the vast majority of individuals work on separate projects in closed source software. Open source projects like Diffgram offer a bright future of shared access to training data tools regardless of the financial status of the country in which a user is living.

Open source tooling also shatters illusions around what is magic and what is standard. Imagine spending more of your precious budget for a database vendor that promises 10× faster queries, only to find out all they do is inject extra indexes. Now in some cases, that could have value, but you would at least want to know up front that you were paying for the concept of indexes, rather than ease of use! Similarly, training data concepts like pre-labeling, interactive annotation, streaming workflows, etc. are brought to the forefront with open source tooling.

Realizing the Need for Dedicated Tooling

As an industry, when we first started working with training data, there was a rush to just "get it done" to start training models.

The questions we considered were along the lines of, "What is the most minimal bare-bones UI for a human to jam annotations on top of data and then into a format a model could use?" This is when people first started to realize the power of modern machine learning methods and just wanted to see "Will this work?" "Can it do that?" "Wow!"

The problems came swiftly. What happens when we move the project from research to staging, or even production? What happens when the annotator is not the same person writing the code, or even located in the same country? What happens when there are hundreds or even thousands of people annotating?

At this point, people often start to realize they need some form of dedicated tooling. Early versions of training data tools answered some of these questions, allowing remote work, some degree of workflow optimization, and scaling capabilities. Quickly, however, more questions entered the picture as pressure on the system increased.

More usage, more demands

To put this very plainly, the moment you have a significant amount of people spending their full working days, eight hours a day, every day, working in a tool environment, everyone's expectations, and the pressure, increases.

Iterative model development, for example pre-labeling, puts pressure on personnel to continually improve training data. While this is desirable, it puts increased pressure on the tooling as well as the people. The more often automation approaches are used, the more the pressure increases. Static pre-labels, explained in Chapter 8, are just the tip of the iceberg. Some automations require interactions, requiring additional interactions between data science, annotators, and annotation tooling.

Many features have been added to address these needs. As tooling providers added more features, the ability to have a smooth workflow became a new issue. Too many features resulted in too many degrees of freedom, and hence, confusion. In response, developers feel a responsibility to limit the degrees of freedom.

Advent of new standards

Tool providers have now had some years of experience and learned many things, from newly constructed concepts specific to training data, to multitudinous implementation details. The off-the-shelf tools that have been developed make the overwhelming into something manageable. This enables you to use these new standards and work at a level of abstraction that's relevant to you and your project.

Yes, we are at the early stages of standard training data. As a community, we are developing everything from conceptual ideas like the schema, to expected annotation functions, to data formats. There is some agreement on what the scope of training data tools is and what standard features are, but there is still a way to go. While there is naturally some overlap, most of the functional areas have differences depending on the media type. For example, automations for text, 3D, and images are all different.

The realization here is that bespoke "Rube Goldberg" machines may answer some of the complexities, but fail to cover the vast space needed. In addition to an academic interest in history, there are practical reasons to inform ourselves about the fundamentals of training data. As someone making a decision today, the context of the progression helps ground where the value is coming from.

I like to think of this as the 30,000-foot view. So if you are thinking about an automation improvement, it's worth reflecting on whether it will apply to all of the media types that are relevant to you. Stepping back to gain a broader perspective can serve as a reminder that any weakness in one area, no matter how insignificant it seems, is likely to create a bottleneck. If it's difficult to get the data in and out, the value of a great annotation workflow is diminished.

Where are you in your journey of needs? Do you already see the need for dedicated tools? For the best quality tools you can get? For a suite that covers the vast training data space?

We all like familiar things. In the same way that office suites offer a familiar environment with clear norms, and software suites offer a similar set of expectations and experiences, from the UI to naming conventions, training data platforms aim to do the same—create familiar experiences in multiple formats, be it text or images.

Naturally, at any given moment, a single team may be focused on a specific data type or types (multimodal). Having familiarity with their work environment helps workers well beyond accomplishing tasks based on their knowledge of similar tasks. New people joining the team can more quickly get up to speed, shared resources can go between projects more easily, and more.

Generally, the decision-making processes about tooling follow a progression:

1. Realizing the need for dedicated tooling
2. Realizing that the complexity of the technical space requires the best possible tooling—not just anything
3. Realizing that the complexity of the user space requires familiarity and shared understanding

As I will explain in more detail in Chapter 7, if you are considering having a director of training data position established, having familiar tooling is of critical importance to this team. The same annotator may easily shift between multiple types of media and projects. This flexibility also helps address the wide scope of data science concerns.

Having a suite does not mean you have an "all-in-one" solution for everything. Data science may have its own suite of tools for training, serving, etc., but that does not preclude point solutions for specific areas of interest. It's more like an order of operations idea—we want to start with the biggest operation, the main suite, and then supplement it where required.

Summary

You now have a roadmap and a general understanding of the process to set up your training data system. From installation, annotation, and embedding through to ML workflows and optimizations. I provided a brief overview of training data tools and then discussed trade-offs and considerations in more depth.

Do you feel confident that you have a grasp of the different concerns among small, medium, and large projects? If not, I'd encourage you to review "Trade-Offs" on page 41 and "Scale" on page 44 before reading on. Training data approaches can vary quite dramatically depending on the project size, so it's important to frame your current learning goal up front.

Finally, as you can see from the history section, training data tools have come a long way, and we continue to see improvements in standards. Getting tools well set up is the practical implementation of thinking about the schema, raw data, quality, integrations, and the human role.

Now that you have an overview of the set-up process, available tools, and trade-offs, it's time to take a deep dive into the schema. What are labels and attributes? What are spatial representations? How can we implement schemas in our training data tools? What is the relation of schema to ML tasks and raw data? Find out in the next chapter.

CHAPTER 3

Schema

Schema Deep Dive Introduction

What do you want your AI system to do? How will it accomplish this? What methods are you going to use?

In this chapter, I dive into some of the foundational concepts around the *schema*, a map between human meaning and machine learning.

The real world is messy. Commercial applications require a level of detail that's hyper-domain-specific. There are many ways to structure all this complexity. In general, these structures are defined in the schema. Further, the schema provides "pivot points" to adapt and change subcomponents over time to better fit current needs.

The schema is important to get right because the rest of the system, including raw data, is defined in relation to the schema.

The schema is the paradigm for encoding all of your commercial knowledge. This can broadly be thought of as *labels* and *attributes* (what something is), *spatial representations* (where something is), their relations to each other, and their relations to external concepts (e.g., series, time). An effective schema relates well to your business needs and your raw data.

More generally, schema is the overall representation of labels, attributes, and spatial information, and their relations to each other. It's how we think about and represent the meaning of what something is, where it is, and more. This builds on the high-level concepts of labels and attributes introduced in Chapter 1. After, I will map these training data concepts back to machine learning tasks.

In this chapter, you will learn:

- Mental models to set up your first schema
- An overview of the landscape of expansion directions for your schema
- Trade-offs of popular methods and tasks
- Specifics of some high-level ideas from Chapter 1

Let's dive into schemas!

Labels and Attributes—What Is It?

Labels and attributes define the "what" something is, what the meaning of the raw data is on a human level. That is, "what" we care about and want the system to learn. In a commercial context, they define what raw data means to our business. In this section, I'll discuss how they relate to other concepts, like spatial types. Using labels and attributes is defining and mapping human meaning in a structured way, and mapping that meaning to technical terms. Together, labels and attributes form the heart of your schema.

What Do We Care About?

Generally, we care *where* something is, *what* it is, and *how* it relates to other things.

Labels and attributes are the tools we use to express "what" something is. In the next section, I will introduce spatial types to discuss "where" something is.

The concept of representing what something is can expand with near-infinite complexity, whereas the spatial location aspects generally have more obvious limits to their expansion. In other words, getting the "what" right is a greater ongoing challenge than understanding the mechanical specifics of "where" something is in a document or image.

Training data schemas should be treated similarly to database schemas (e.g., like Postgres schemas with tables, views, data types, etc.).

Introduction to Labels

Labels are the "top level" of semantic meaning. In the base case, they represent only themselves. In most cases, though, labels organize a set of attributes.

To help ground this idea, although it's an imperfect analogy, I compare it to SQL, as you can see in Table 3-1.

Table 3-1. Training data schema relational concept and analogy to SQL

Concept	Training data	SQL
Attribute	Attribute	Column (attribute)
Set (of attributes)	Label	Table
Schema (set of sets)	Set of labels	Set of tables (and other objects)

In SQL, each column has a type, and in training data, each attribute has a type. In SQL, a table organizes multiple columns, while in training data, a label organizes multiple attributes. In training data, attributes may be shared between labels, which is roughly analogous to foreign keys.

In E.F. Codd's *The Relational Model for Database Management*, he mentions that *columns* were originally thought of as *attributes*.[1] While far from a perfect analogy, it helps convey the general idea.

When an end user is annotating, organizing sets of attributes into labels also can help hide irrelevant options. For videos, labels help constrain relationships and organize sequences.

It's expected that some of the specific organizational principles discussed here will be implementation specific and change over time. In general, the broad strokes will be similar. As this new area of training data continues to be refined, standards will continue to evolve.

Next, we will talk about attributes, which is where the bulk of the schema definition usually lives.

Attributes Introduction

Attributes represent the bulk of the "What is it?" which is the heart of the human-encoded meaning and the technical definition of data. Attributes are usually defined to include, at minimum, the following structures: the human question or prompt, the form type, and technical constraints. This set of human and machine definitions together make one "attribute."

Training data attributes may appear superficially simple or similar to other technologies; however, in practice, there is a lot of complexity at this intersection of both human and machine-centered definitions.

---

1 "In subsequent papers (e.g., Codd 1971a, 1971b, and 1974a), I realized the need to make this distinction, and introduced domains as declared data types, and attributes (now often called columns) as declared specific uses of domains." E.F. Codd, *The Relational Model for Database Management*, Version 2 (*https://oreil.ly/DGq4x*) (Addison-Wesley, 1990), 43.

In the same way that training data is a combination of raw data and human-defined meaning, attributes are a combination of technical definitions and human-centered definitions. In order to be useful for machine learning, both the technical definition and the human definition are needed. Attributes are the joint representation of those two things.

More technically speaking, attributes can be thought of as well-defined forms, or as "data classes meet UI specifications." One way to wrap your head around this is to think of a spectrum between forms and classes and put attributes somewhere on that spectrum. A form can be arbitrarily complex, but usually isn't thought of as having defined types, like a class. Further, while the implementation of a form may have validation, it's usually end-user validation, not a formal database constraint. Because the training data is relied upon by the ML program, and is usually expected to be queryable, attributes have more "structure" than a typical form. Conversely, due to the expectations of human control, attributes usually have a flair of "form-like" behavior, more than a typical programming class or database table definition would have.

In practice, attributes fill a need for training data that is distinctly different from other technologies. As this area continues to evolve, I expect that attributes will continue to expand.

Attribute concepts

The following concepts are generally present in all forms of schemas:

Relation to annotations
A single annotation may have no attributes, a single attribute, or multiple attributes.

Scope
An attribute may be scoped to a single annotation, a single file (e.g., an image), or a whole business concept (e.g., a driver's license).

The question for humans
A question for a human to consider. For example, "Is this person happy?" A human will be selecting these values and/or reviewing them. In that context, usually, each group will have additional information, for example, about the prompt, display, or order.

Form type
Examples of form types include tree, multiple select, select, text, child group, date, etc. Their use in training data straddles that line between user interface form types and data types. For example, a slider UI control could collect float or integer data. In the most general case, attributes are often treated as strings. Where needed, both a UI type and a data type can be stated.

Constraints (or bounds)

There may be constraints with the form collection; for example, how many are allowed to be selected. Continuing the slider example, it would have a lower and upper bound.

Predefined selection

In this instance, an administrator generally defines valid values.

Templates

Usually, attributes are defined as some kind of template. Values are unique to each instance, and may be concrete or a reference.

Typically, any kind of number collection, free text, date, etc. would be concrete, e.g., "3.14." However, a reference ID can be used if a known set is provided for, say, a selection from a list of six elements.

Examples of generic attributes:

- Occlusion (blocked/out of view) and truncation (out of frame), as shown in Figure 3-1.

- Depth/hierarchy of labels/meaning (e.g., the type of action moving, jumping, running).

- Directional vector (e.g., front/back/side).

Figure 3-1. Example of occlusion and truncation

To make this easier, we use constraints. We may set a constraint, for example, that a car can have a directional vector but a sidewalk does not. A disease may have multiple types, or we may constrain the system to only choose a single type. In the simplest case, if we imagine the labels "cat" and "dog" being the two options, we are constrained to only those two choices.

Schema complexity trade-off

The complexity of the schema affects the ML program and human supervision. The overall schema may be different from the supervision that is shown to users at any given moment in time. In that sense, the ML predictions and the end-user correction may be decoupled. Figure 3-2 shows an example of the complexity and performance trade-offs you need to consider.

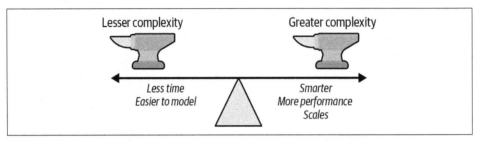

Figure 3-2. Spectrum of common schema trade-off considerations

The benefit of more schema is often a "smarter" system. For example, if we don't have labeled data on "is offensive?" then we wouldn't be able to train a model. More labels also offer more insight into performance. As will be explained in upcoming "Using Spatial Types to Prevent Social Bias" on page 78, a higher level of schema is required to prevent social bias.

Systems tend to expand how many labels and attributes they have over time. A good "back of the napkin" way to understand how complex the schema is involves multiplying the number of attributes by the number of annotations.

The media type may affect the complexity trade-off greatly. For example, with images, if every object needs to be labeled, then the complexity of the schema may be multiplied by the volume of objects per image. In a video context, this may be further multiplied by each frame. The more complex the schema is, the more complexity is there for model training.

Attribute depth

Imagine a grocery store checkout system. There may be many sizes of a specific brand of cereal box, and we aim to identify the specific stock keeping units (SKUs), the codes used to keep track of inventory, for the different sizes. In this context, we may wish to "database back" our attributes. If a particular size is no longer sold, the options presented to the cashier can change to reflect that.

Much further than that example, we can load thousands or even millions of options. Putting aside UI challenges regarding searching and selecting each attribute, this can be a perfectly reasonable solution, depending on your use case. At the time of writing, there is no "right" answer here. A system may easily have a handful of attributes, or thousands.

Schema depth may also be affected by conditionals and complex hierarchies. At any point, we can expand an attribute into a child node or a list of nodes. For example, selecting "No" could expand a "No—Options" node. It is possible for any given selection to expand into a child node. Generally, the same principles discussed here apply to these deeper layers of the system.

Attribute Complexity Exceeds Spatial Complexity

Earlier, I mentioned briefly why "What" data is important. To expand on this, consider that oftentimes, there are severe limits to the definition of "where." Note that "where" (localization or location) will be covered in more detail in "Spatial Representation—Where Is It?" on page 78, which is focused on the contrast between the two.

In complex cases, multiple sensors can be combined to represent a complex 3D view of where. In segmentation cases, location can be defined at the per-pixel level. However, in each of these cases, there's a relatively finite, limited ability to declare spatial locations. There are only so many pixels, or voxels, etc. It's like a slot, either there's an object in the slot, or there is not. Further, often in practice the localization is only a small part of the overall picture. Consider the sports player, as shown in Figure 3-3. Sure, we can say the body or even a specific limb is in a certain position, but what does that really mean at the level of playing soccer? Is the player "attacking" or "defending"? Do they have "possession" of the ball? What about identifying who the player is?

Figure 3-3. Example of a general spatial localization with a bounding box

Further, through a variety of assisted and pixel-based methods, localization is, relatively speaking, easier to solve. In contrast, the "meaning" of some concepts in life are the subject of philosophy books.

To be clear, there can still be significant challenges in localizing novel objects, but generally, for ongoing concerns, the "meaning" of what they are becomes more central.

The hidden background case

When we think about where something is, it's sometimes more useful to start with where something is *not*. For example in the image context, when we say "it's a traffic light," at some point, what we are actually saying is that 96% of the image is *not* a traffic light. Meaning that by defining a single traffic light, we are also assuming it will be predicting the "background" (not traffic light) class. It is hidden because it's often implicitly assumed and not directly labeled as "background."

Continuing the example, one way to detect foreground objects from the background is the concept of an "objectness" score.[2] This is a concept of detecting that an object exists, distinct from the background, without knowing the class of the object. It would be difficult to detect the traffic light as an object without having both a background class and some concept of generic objectness (location). So while theoretically the spatial location of the traffic light, the label "traffic light" (the what), and the implicit background class are all different things. At implementation time they start to blend together, e.g., through the implementation of the objectness score to detect objects relative to a background. More broadly, I surface this simply to be aware that training data is created without a one-to-one correlation to how it will be consumed by a data science model.

As another example, it may be visually easier for a hierarchical "nested" list to be displayed to a human. But in actual practice, that nesting may be implemented in the ML network in many ways. For example, the system could "flatten" the labels (red_occluded_20, red_occluded_40), by combining multiple models, or use an architecture that supports nesting.

Depending on the implementation, it's possible to predict that a generic object (or segmentation mask) exists and what that object or segmentation is. These methods may change in the future, and there is already a significant breadth of approaches.

Example of sharing attributes between labels

Earlier, I introduced the label as the highest level of semantic meaning, defining "what" something is, for example, a "strawberry" or "leaf." Attributes are introduced as the breadth and depth of "what." So how do these work together?

Let's imagine we are building a system to automatically detect what percent of fans at a sports match are cheering for one team over another. And perhaps it also needs to identify "offensive" content.

We may want to identify articles of clothing, such as T-shirts, pants, and ballcaps. There are "what" representations like color, team logos, is it offensive, etc. that are

---

2 The implementation of that is beyond the scope of this book.

common to all of those items. There may also be certain things that are only relevant to, say T-shirts or ballcaps.

One way to structure this is to have T-shirts, pants, and ballcaps as the labels. Then, you can create an attribute called "Color" with the various properties, such as "Red," "Blue," "Green." All of the labels can have access to this attribute.[3] This is illustrated in Figure 3-4.

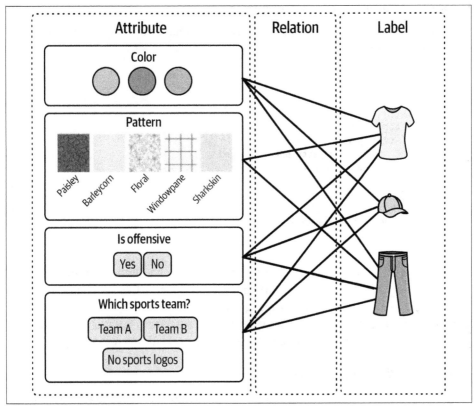

Figure 3-4. Relationship between attributes and labels

There are lots of trade-offs to think about here. Is it worth it to identify color? Perhaps it's common to wear a team's colors, but a logo may not be visible. So perhaps having a good idea of color may be used for further downstream processing to determine which team the person is cheering for.

---

3 Technically: An attribute may be represented in multiple ways, e.g., as a "group" with "attributes" as children, or as an "attribute" with "properties" as children.

Technical Overview

This section is an introduction to some of the more engineering-focused ways of representing attributes. We'll look at common data structures for them, types, and the relation of these engineering concepts to UI-centric user concepts. For the specific implementation details, see the docs of your training data tools.

Example of an attribute in relation to an instance

Consider the following code snippet, where a minimal group may be represented in the following form:

```
"instance_list": [
    {
      "type": "box",
      ...
      "attribute_groups": {
        id_of_group : {
          "id": id_of_attribute_selected
          "kind": "select"
        }
      }
    },
    {... another instance }
  ...
  ]
```

In this example, there is a key `instance_list`, where each annotation is one instance. Each instance has a key `attribute_groups`, the preceding defined attributes. So, for example, if there were three groups, there would be three keys in the `attribute_groups` dictionary. Each attribute then has information about its selected states, such as IDs of value selected, or the raw characters of free text, etc.

Data representations for engineering

Your schema will likely change over time. Therefore, for ease of change management, it's sometimes useful to either "lock" the schema or pass by value, even when it's possible to pass by reference. At a very high level, it's easier to think of labels as "slow-moving" concepts and attributes as "fast moving." It's relatively easy to add and remove attributes from labels, but changing labels may introduce breaking changes to other attributes. The analogy of tables (labels) and columns (attributes) in a database works well here. Chapter 4 will discuss related topics in more depth.

Examples of attributes

Attributes represent the union between human-defined meaning and machine learning readable data.

In practice, training data software has a number of implementation details to make that a reality. As a high-level overview, here are some examples of common attributes, their high-level types, and constraints.

As shown in Table 3-2, some things in a schema can be defined as references. For example, here, each option could be represented as a single radio select. However, for other types of attributes, like free text, the answer may be unique for each instance.

Table 3-2. Table of technical specifications of attributes

Attribute	Example	Output types	Store by reference or value	Constraints
Select (drop down) or radio buttons (meaning all options shown)		String	Reference	Allow list
Multiple select		List	References	Allow list
Free entry		String, integer	Value	Blocklist
Slider				Force type (string, integer, float, etc.)
				Force range
Date		Date string or ISO 8601	Value	Is date
Child node	Expands another attribute group	String, e.g., unique hash; Integer, e.g., an integer primary key	Reference	

Technical example of an attribute

Now that we've walked through the basics of attributes, let's look at a brief example.

The following code snippet is a quick preview that assumes that some kind of map of `id_of_group` to information about the group is available, such as:

```
"attribute_groups_reference": [
    {
      "id": id_of_group ,
      "options": [
  {
          "id": id_of_attribute_selected
          "name": "option_name"
        }
    ],
      "prompt": "user defined prompt",
    }
  ]
```

In practice, there are multiple groups, with multiple types (as shown in Table 3-2), and each type will have a different format.

Spatial Representation—Where Is It?

In many cases, we need to know where an example is. For example, in an image, we may need a bounding box to know where the object is. In text, we need start and end characters or tokens. In audio, we need start and end timeslices. Check the docs of your training data tooling for the specific representations.

This section will focus on more conceptual-level items and ground them with some specific examples around computer vision. Later in the chapter, I'll carry the computer vision example types forward by relating them to ML tasks. Note that spatial representations are also known as localization, locations, shapes, and drawing tools.

I'll cover three things specifically:

- Using spatial types to prevent social bias
- Trade-offs with types
- Computer vision spatial types in detail

Using Spatial Types to Prevent Social Bias

Imagine that for our sports fan detector, we have a "Person" as the top-level object. And then, for detecting if they are wearing an offensive shirt or doing an offensive action, we have an `is_offensive` attribute. And let's assume the annotation instance has one spatial type, a bounding box for the Person, shown in Figure 3-5.

Figure 3-5. Example of schema with bounding box type and attribute "Is Offensive" describing a whole person

While we, as humans, are thinking about the shirt, the actual annotated data is of the *whole person*. This is risky, because what we have actually encoded into the machine is that the whole area is offensive. Not what we, as humans, know immediately, that it's just the shirt that's the offensive part.

To help visualize this, consider Figure 3-6. The left and right images show that we are declaring to the model that the top and bottom half of the image are *equally* offensive.

Figure 3-6. How a machine may read the bounding box

As humans, we know the offensive part is the T-shirt, not the person's face. To the machine, the top and bottom are declared to be equally offensive, which may lead to it training on human facial features falsely.

While it may appear reasonable to think we can label the whole image and trust the model to figure it out, this is usually a faulty assumption. This is because at training time, since the entire image was declared to represent "offensive," the ML program may make predictions based on aspects of the human face, and not the T-shirt, as may be assumed. If the images happen to have more of a certain ethnic group, that may affect the model result, leading to very real, unexpected, and likely undesired outcomes. Worse yet, since the validation or test sets could easily contain this bias, it's possible the model could get a strong technical score, while still containing the bad social bias.

For example, if, by chance, in the data, racial group A had "offensive" T-shirts on 65% of the time, while group B had them on 35% of the time, it's likely that the model will learn to associate racial features of group A with offensive. This is because the model may just as easily use a feature from the racial features of the face instead of the shirt, therefore falsely classifying someone who has some facial or ethnic properties as "offensive." Remember the concern here is not if it can "work," or learn the offensive concept, but if it's learning it in a way that reflects our social goals, such as focusing on the T-shirt itself and not the skin color of the person wearing it.

In this example, the relationship is slightly more obvious with labels because they are "top-level," but it can be more subtle in spatial locations or "buried" groups of attributes. This same issue can happen when you have tens or hundreds of attribute groups and thousands of sub items. Similarly to how a classic program must specify instructions very clearly, an ML program must have a very clear schema to avoid unwanted results.

Consider that most systems and people will use this type of training data without looking at any (or any material volume of) the samples, and most likely without critically analyzing its schema. To avoid this, we want to make sure the schema encodes what's actually present, as accurately as reasonably possible, and when possible that we record assumptions made around the system.

One way to avoid spatial bias

Imagine you are teaching a child about what kinds of clothing are appropriate to wear in public. You might say something like, "That person is wearing an offensive shirt." You wouldn't say, "That person is offensive." While the latter may be true, it's not an accurate way to describe the logical situation as observed in the image. You can see this represented in Figure 3-7.

One solution to this is to break out the label into two parts. Then the spatial location (e.g., here, of the dashed-line box) can better represent the label. The key point here is that we are trying to get the spatial location to be representative of the item of interest—and only the item of interest.

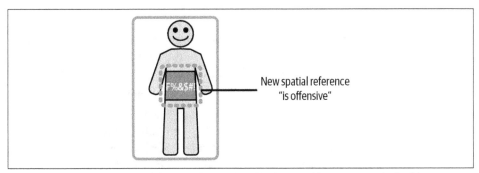

Figure 3-7. Avoiding spatial bias with newly added spatial location in dashed lines

We can generalize this concept to state a rule of thumb: "The spatial location should match the concept of concern as closely as is reasonably possible."

Theoretically, in the future, it should be possible for a model to learn a reasonable representation based off of higher-level annotations such as the whole image. However, while perhaps future training methods will allow for preventing or mitigating biases like this automatically, at the moment there is not a strong industry consensus on this, and it does not happen by default in popular image models at the time of writing.

In the meantime, we can relatively easily mitigate this risk at the training data level simply by being aware of it and being more precise in how we lay out our classes and training data. This layout need not be taken to extremes; for example, I'm not suggesting segmenting every pixel of every letter, because the risk of it learning a shirt pattern as a bad bias seems low. But I do recommend that, at least, the broad, obvious cases should be considered when there is a clear bias risk.

To further appreciate the importance of this issue, consider another example. Imagine an alarm system that is monitoring for a gun, knife, bomb or some other real-time threat (such as airport security). If the system is trained on images of people with guns, but doesn't separate the person from the gun, there is a risk of it falsely triggering when a person of a certain ethnic background is present (when no gun is!). Victims of biases like this have their time wasted and are subjected to unnecessary scrutiny, at the very least, and the consequences can be far more grave. Meanwhile, the security personnel's time is also being wasted, as are the organization's resources.

If the spatial area annotated is of the true object we care about, in theory, all of the background data should not be relevant. In practice, this may not be true, however, so awareness of the overall data being used is still a joint responsibility.

Joint responsibility

If you are a data specialist, you may think this doesn't apply to you. Because of course, this is a "schema-level" concern. Equally, if you are designing the schema, you may assume that the data specialist will notice it and raise an alarm. Clearly, it is a *joint* responsibility to identify when the schema is at risk of creating bad data.

Another reason to be aware of the actual data and data's relationship to the schema is to create better-quality data, which results in faster and better models. For example, here you may need a lot more data to identify offensive items if you are doing it at the person level. By scoping the spatial location to better match the object in question, we improve the overall quality of the data.

As mentioned earlier, this is not an ethics book. Here, I outlined the direct technical effects of choosing attributes and their related spatial locations. There are many more trade-offs, and I encourage you to think of this as just an introduction to new reasoning about the schema, and who is responsible for it.

Trade-Offs with Types

Choosing a great spatial representation depends on a number of factors, including the task at hand, performance and bias concerns, and the kind of data being annotated.

As mentioned earlier, I'll focus in on computer vision for the sake of illustrating the trade-offs. Similar trade-offs exist for other types, such per-token spans versus per-character spans, using or not using relationships in text, etc. Figure 3-8 walks through each of these trade-offs as they relate to classification, object detection, and instance segmentation.

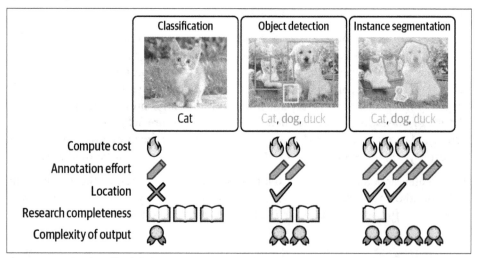

Figure 3-8. Trade-offs of different spatial types for computer vision tasks (Adapted from original illustration by Anthony Sarkis)

As with any solution, there are trade-offs that should be considered. These include:

- *Relevancy to task.* In general, the type must "match" the task or concept to some extent.
- *Complexity of ML output.* For data engineering and applications. For example, even if you could magically annotate and predict a polygon, keypoint, or more complex shape, can your application actually make good use of that output?
- *Compute cost.* More complex annotations require more effort to predict in the ML model, and may require more compute cost to pre-label, too.
- *Spatial location value needed.* Do you need a segmentation-level prediction?
- *Research completeness.* (ease of modeling).
- *Annotation effort.* Of course, the more complex the annotation, the more time it takes. But often it's in a step function type fashion, e.g., segmentation may take 10 times longer to annotate than a box.

Similar trade-offs exist as more precision is added in other media types. For example, document > sentence > word > character.

While research continues to make segmentation easier and easier, e.g., the Segment Anything Model (SAM) and similar, in practice there are still limitations and considerations. At the time of writing, it still remains generally easier to predict a box, and SAM still requires a degree of prompting or fine-tuning. To repeat the point of complexity of output, sometimes a segmented mask is actually undesirable; it's not automatically better.

Computer Vision Spatial Type Examples

Let's ground the above with some specific examples from computer vision. These types are listed to give examples, and are not a comprehensive resource. As mentioned earlier, other media types have other spatial representations.

For computer vision the three most popular spatial types are polygon (for segmentation), box, and full image.

Full image tag

Full image tags lack a spatial location, which in some cases may limit their usefulness. In cases where the schema is very broad, tags can still be very useful. However, in other cases, if we don't know where the item is, or there are multiple items, the value is less. In those cases, there is a faulty assumption that the "item" of interest already fills the frame, which may not be realistic or may overly constrain potential results.

Box (2D)

The bounding box is one of the most "tried and true" methods of object detection.

A box is defined by exactly two points. Only two pieces of information are needed to store a box: it can be defined as either the top left point (x,y) and (width, height), or as an (x_min, y_min) and (x_max, y_max) pair of points.[4] A box may be rotated and defined by an origin and rotation value.

Polygon

A polygon is naturally defined by at least three points. There is no predefined limit to how many points a polygon may have. Often, points are created via an assisted process, such as dragging the mouse or holding the shift key, or a "magic wand"–type tool. A polygon may be drawn around the border, or a polygon tool may be used to correct (e.g., add or subtract from) a previously predicted dense mask. A single annotation example may have multiple sets of polygons, e.g., to indicate "holes" or other related spatial features.

Ellipse and circle

Ellipses and circles are both defined by a center point and a radius (x,y). They are used to represent circular and oval objects, and they function similarly to a box. They can also be rotated. Figure 3-9 shows two examples of this.

Figure 3-9. Examples of ellipses and circle

---

4 Anthony Sarkis, "How Do I Design a Visual Deep Learning System? An Introductory Exploration for Curious People" (*https://oreil.ly/PtRMs*), *DiffGram–Medium*, March 4, 2019.

Cuboid

A cuboid has two "faces." This is a projection of a 3D cuboid onto 2D space. Each face is essentially a "box," as shown in Figure 3-10.

Figure 3-10. Example of cuboid

Types with multiple uses

Some types have multiple uses. A polygon can be easily used as a box by finding the most extreme points. There is not exactly a "hierarchy."

Other types

There are other types such as lines, curves, quadratic curves, and others.

Raster mask

Every pixel may be labeled with a specific class. "Brush" type tools allow for annotating many pixels at once. The trade-offs of raster masks relative to polygons are contextual. Usually, a raster is less desirable to annotate with; however, it can be relevant in cases where automatic tools are used or a polygon would simply be incapable of capturing assumptions in the data.

Polygons and raster masks

There are various algorithms that can transform between polygons and raster masks. For example, a polygon can be turned into a pixel-wise mask and a dense mask can be approximated into a polygon using a curve decimation algorithm like Ramer–Douglas–Peucker. In practice, the end goal of prediction is similar, so either a polygon or a raster mask can be thought of as the "highest" level of 2D localization.

Keypoint geometry

One of the spatial types that's not predefined is the "keypoint" spatial type. In this case, you can use an interface to construct the spatial template you desire. A keypoint can be used for pose estimation, known geometry that doesn't conform to typical structures, etc. The keypoint type is a graph with defined nodes (points) and edges. For example, a keypoint could depict an anatomical model of a person, with fingers, arms, etc. defined, as shown in Figure 3-11.

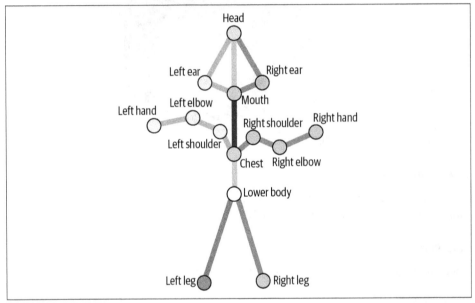

Figure 3-11. Keypoint geometry example

Custom spatial templates

You can also create default spatial templates. These can be both for custom shapes, or essentially another path to keypoints. The main difference is that keypoints often also include a referenced relation or edge to each point, whereas a generic polygon template, like the diamond shown in Figure 3-12 (left) does not specifically enforce this.

Figure 3-12 shows visual examples of an instance template shaped like a modified diamond, as well as the UI showing the process of creating a new template.

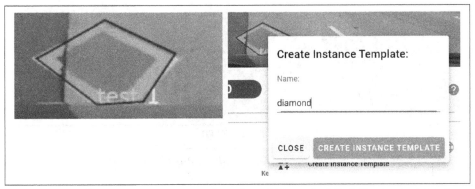

Figure 3-12. Left, custom instance template shaped like a modified diamond; right, the UI showing the process of creating a new template

Complex spatial types

A complex type refers to a spatial location defined by a set of multiple "primitive" types, such as box and polygon, as defined above. This is similar to, but different from, the concept called "multimodal." Generally, multimodal refers to multiple items of raw data. A common use case for complex types is complex polygons, such as a polygon that is partially defined by segments of quadratic curves and partially by segments of "straight" points.

Relationships, Sequences, Time Series: When Is It?

Many of the most interesting cases have some form of relationship between instances.

Sequences and Relationships

Imagine a soccer game. A soccer ball is touched by a player. This could be considered an "event" taking place at a given frame, e.g., frame 5. In this case, we have two instances, one of a "Ball" and one of a "Player." Both of these are in the same frame. They can be formally linked with an annotation relationship between the two instances, or can be assumed to be linked in time because they occur in the same frame.

When

Another big concept at play here is relationships. Consider upgrading our still image example to a video context. Take a look at Figure 3-13. From the human perspective, we know the "car" in frame 1 is the same car as the one in frame 5, 10, etc. It's a persistent object in our minds. This needs to be represented in some form, often called a sequence or series. Or, it can be used to "re-ID" an object in a more global context, such as the same person appearing in different contexts.

| Frame 1 | Frames 2 through 9 | Frame 10 |

Figure 3-13. Example showing the same car with multiple backgrounds

Guides and Instructions

When we're working with training data, we spend a lot of time getting the model to be able to identify images using labels, like the one shown in Figure 3-14. Of course, how do we know to label it "liver" at all? I sure wouldn't be able to figure that out without some form of guide. What about sections of the liver that are occluded by the gallbladder or stomach? Figure 3-14 provides an example of this anatomical guide and potential for occlusion.

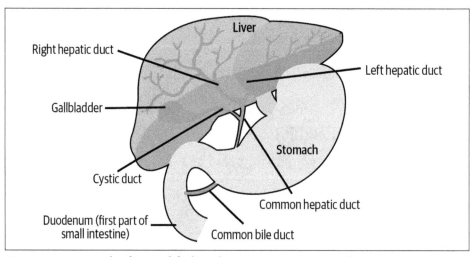

Figure 3-14. Example of a simplified single-image annotation guide; in practice, these guides are often very long and contain many examples and are made with the real media being used (e.g., images or videos)

In a training data context, we formalize the concept of a guide. In a strict sense, the guide should be maintained with as much respect as the actual training data, because the "meaning" of the training data is defined by the guide. In a sense, guides address the "How" and "Why."

We have covered some of the high-level mechanics of representing training data. While complex scenarios and constraints can be a challenge, often the "real" challenge is in defining useful instructions. For example, the NuScenes' dataset has about a paragraph of text, bullet points, and 5+ examples for each top-level class.

In Figure 3-15, the example image shows the difference between a "bike rack" and a "bicycle." This would be provided as part of a larger guide to annotators.

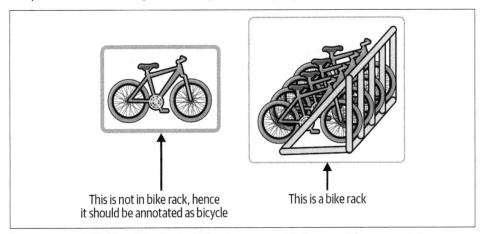

This is not in bike rack, hence it should be annotated as bicycle

This is a bike rack

Figure 3-15. Example of single clarifying image provided in a guide to annotators

Judgment Calls

To see how some of this can quickly get dicy, consider "drivable surface" versus "debris." NuScenes defines it as "Debris or movable object that is left on the driveable surface that is too large to be driven over safely, e.g., tree branch, full trash bag etc."[5]

Of course, what's safe to drive over in a semitrailer versus a car is different. And this isn't even getting into choice semantics like "should you drive over debris to avoid rear-ending someone?"

Relation of Machine Learning Tasks to Training Data

Training data is used in a machine learning system. Therefore, it's natural to want to understand what ML tasks are common and how they relate to training data.

There is general community consensus on some of these groupings of tasks. There are many other resources that provide a deeper look at these tasks from the machine learning perspective. Here, I will provide a brief introduction to each task from the training data perspective. To start, we'll use an example image, shown in Figure 3-16.

---

5 See the NuScenes annotator instructions on GitHub (*https://oreil.ly/LaDOt*).

Semantic Segmentation

In semantic segmentation, every pixel is assigned a label, as shown in Figure 3-16. An upgraded version of this is called "instance segmentation," where multiple objects that would otherwise be assigned the same label are differentiated. For example, if there are three people, each person would be identified as different.

Figure 3-16. Example of a segmented image where each pixel is assigned a label (shown as different shades)

With training data, this can be done through "vector" methods (e.g., polygons) or "raster" methods (picture a paint brush). At the time of writing, the trend seems to favor vector methods, but it is still an open question. From a technical perspective, vector is much more space efficient. Keep in mind that the user interface representation here may differ from how it's stored. For example, a UI may present a "bucket" type cursor to select a region, but still represent the region as a vector.

Another option question at the time of writing is how to actually use this data. Some new approaches predict polygon points, whereas the "classic" approach here is per pixel. If a polygon is used for training data and the ML approach is classic, then the polygon must go through a process to be converted into a "dense" mask. (This is just another way of saying a class for each pixel.) The inverse is also true if a model predicts a dense mask but the UI requires polygons for a user to more easily edit.

Note that if a model predicts vectors—polygons—to fulfill the classic definition of segmentation, then the vector must be rasterized. Rasterized means converted into a dense pixel mask. Depending on your use case, this may not be necessary. Keep in mind that while a per-pixel mask may appear more accurate, if a model based on a vector approach more accurately captures the relative aspects, that accuracy may be more illusionary than real. This is especially relevant to contexts that contain well-documented curves that can be modeled by a few points. For example, if your goal is to get a useful curve, predicting the handful of values for a quadratic curve directly may be more accurate then going from per-pixel predictions back to a curve.

While there are many great resources available, one of the most prolific is the site Papers with Code, which lists over 885 computer vision tasks, 312 NLP tasks, and 111 in other categories.[6]

Let's walk through an example of a medical use case. If one has a goal of doing automatic tumor segmentation (e.g., of CT scans), segmentation training data is needed.[7] Researchers routinely note the importance of training data. "The success of semantic segmentation of medical images is contingent on the availability of high-quality labeled medical image data."[8]

Training data is created in conjunction with the task goal, in this case, automatic tumor segmentation. In the example depicted in Figure 3-17, each pixel corresponds to a classification, this is known as "pixel-wise" segmentation.

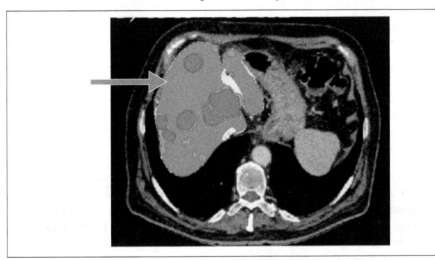

Figure 3-17. Example of tumor segmentation, a subcategory of segmentation (Source: Simpson et al., "A Large Annotated Medical Image Dataset for the Development and Evaluation of Segmentation Algorithms", Figure 1 (https://oreil.ly/gVrYA))

Each shaded pixel (the specific shade indicated by the arrow) is predicted as "liver" on this CT scan (the spatial location where something is, as defined earlier). In the context of training data, this could be annotated with either pixel brushes "raster" formats or with a polygon. The label "liver" is the What, as defined earlier.

---

6 Browse State-of-the-Art (*https://oreil.ly/uxqf3*), accessed Oct. 4, 2020.

7 Patrick Ferdinand Christ et al., "Automatic Liver and Tumor Segmentation of CT and MRI Volumes Using Cascaded Fully Convolutional Neural Networks" (*https://oreil.ly/AO5ff*), arXiv:1702.05970v2 [cs.CV], February 23, 2017.

8 Amber L. Simpson et al., "A Large Annotated Medical Image Dataset for the Development and Evaluation of Segmentation Algorithms" (*https://oreil.ly/gVrYA*), arXiv:1902.09063v1 [cs.CV], February 25, 2019.

Image Classification (Tags)

An image may have many tags. While this is the most generic, consider that all the other methods are essentially built on top of this. From an annotation perspective, this is one of the most straightforward.

Object Detection

Object detection refers to when the model detects the spatial location of multiple objects and classifies them. This is the classic "box drawing." It's generally the fastest spatial location annotation method and provides the big leap of getting spatial location. It's a great "default choice" if you aren't quite sure where to start, because it often provides the best bang for your buck.

While most of the research has centered around "boxes," there is no requirement to do this. There can be many other shapes, like ellipses, etc.

Pose Estimation

At a high level, pose estimation is "complex object detection." Instead of a general shape like a box, we try to get "keypoints" (graphs). There are graph-like relationships among the points, such that, say, the left eye must always be within some bounds of the right eye, etc.

From a training data perspective, this is handled via the keypoint template. For example, you could create an 18-point representation of a human skeleton to indicate pose and orientation. This is different from segmentation/polygons because instead of drawing the outline, we are actually drawing "inside" the shape. Figure 3-18 shows two people with keypoints assigned.

Figure 3-18. Example of people with keypoints (pose) drawn

Relationship of Tasks to Training Data Types

There is not a one-to-one correlation between annotation data and machine learning tasks. However, there is often a loose alignment, and some tasks are not possible without a certain level of spatial data. For example, you can abstract a more complex polygon to a box for object detection, but cannot easily go from a box to segmentation. Table 3-3 provides additional detail.

Table 3-3. Mapping of spatial types to machine learning tasks

Training data spatial type	Machine learning task example
Polygon, Brush	• Segmentation • Object detection[a] • Classification
Box	• Object detection • Classification
Cuboid	• Object detection • 3D projection • Pose estimation • Classification
Tag	• Classification
Keypoints	• Pose estimation • Classification

[a] Well, technically, it can be used for object detection, but usually that's a secondary goal because it's faster to do a box for the generic detection base.

General Concepts

The following concepts apply to, or interact with schema concepts in general.

Instance Concept Refresher

Virtually everything discussed here relates to an instance (annotation). An instance represents a single sample. For example, in Figure 3-19, each person is an instance.

An instance has references to labels and attributes, and may also store concrete values, such as free text that's specific to that instance.

In a video or multiple frame context, an instance uses an additional ID to relate to other instances in different frames. Each frame still has a unique instance, because the data may be different; for example, a person may be standing in one frame and sitting in another, but they are the same person.

To illustrate a subtle difference, consider that here, all three people have the same "class," but are different instances, as shown in Figure 3-19. If we didn't have this instance concept, then we would get a result like the middle image.

Figure 3-19. Comparison of treating people as one group, versus as individual people (instances); adapted from David Stutz (https://oreil.ly/jDMTy)

Consider that for every instance, there can be n number of attributes, which may, in turn, have n number of children, and the children may have arbitrary types/constraints (such as selection, free text, slider, date) of their own. Yes—in fact, each instance is almost like a mini graph of information. And if that wasn't enough, the spatial location may actually be 3D. And there may be a series of frames.

Modern software tools handle the relationships between these concepts well. The challenge is that the data supervisor must, at least to some degree, understand the goal, so as to do a reasonable job.

Further, the types must align with the use case in the neural network. Common network architectures have assumptions to be aware of when constructing training data.

Upgrading Data Over Time

In some cases, it can be quite reasonable to "upgrade" data over time. For example, a common pattern unfolds as follows:

1. Someone runs a "weak" classification to tag images, purely with the intent of identifying "good" training data.

2. Humans create bounding boxes.

3. At a later point in time, a specific class is identified as needing segmentation. The bounding box can then be used for an algorithm designed to "generate" segmentation data once a rough localization has been done (the existing bounding box).

The Boundary Between Modeling and Training Data

There is often a disconnect between the training data and how the machine learning model actually uses it. One example shown earlier is how the spatial locations do not have a one-to-one mapping to machine learning tasks. Some other examples include:

- An ML model uses integers while humans see string labels.
- A model may "see" a dense pixel mask (for segmentation) but humans rarely, if ever, consider single pixels. Often, segmentation masks are actually drawn through polygon tools.
- ML research and training data advance independently of each other.
- The "where versus what" problem: humans often conflate where something is and what it is, but these may be two distinct processes in an ML program.
- The transformation of human-relevant attributes to use in the model may introduce errors. For example, humans viewing a hierarchy, but a model flattening that hierarchy.
- The ML program's efforts at localization are different from those of humans. For example, a human may draw the box using two points, while the machine may predict from a center point, or use other methods that don't have anything to do with those two points drawn.[9]

Basically, what this means is:

- We must constantly remind ourselves that the ML modeling is different from the training data. The model uses integers, not human labels, has a different localization process, different assumptions about where versus what, etc. This difference is expected to widen over time as training data and ML modeling continue to progress independently.
- Capture as many assumptions about how the training set was constructed as possible. This is especially vital if any type of assist method is used to construct the data.
- Ideally, whoever manages the training set should have some idea of how the model actually uses the data. And whoever manages the model should have some idea of how the set is managed.
- It's crucial that training data can be updated in response to changing needs. There is no such thing as a valid static dataset when the ML programs interacting with the training data are changing.

---

9 See SSD and how some of the regression points work here (*https://oreil.ly/sqoeX*).

Raw Data Concepts

Raw data means the literal data—the images, videos, audio clips, text, 3D files, etc. being supervised. More technically, it's the raw bytes or BLOB that is attached to the human-defined meaning.

The raw data is immutable, whereas the human-defined meaning is mutable. We can choose a different schema, or choose how to label it, etc.; however, the raw data is what it is. In practice, the raw data usually undergoes some degree of processing. This processing generates new BLOB artifacts. The intent of this processing is usually to overcome implementation-level details. For example, you might convert a video into a streamable HTML-friendly format, or re-project coordinates of a GeoTIFF file to align with other layers.

These processing steps are usually documented in the training data software. Metadata about the processing should also be available. For example, if a certain token parsing scheme was used in a pre-processing step, the machine learning program may need to know that. Usually, this is not a major issue, but it's something to be aware of.

It's best to keep data as near to the original format as possible. For example, a natively rendered PDF is preferable to an image of a PDF. Native HTML is better than a screenshot. This is complementary to implementation details of generating new artifacts. For example, a video may play back in a video format, and may additionally generate images of the frames as artifacts.

Accuracy is how close a given set of measurements are to their true values. Some ML programs have accuracy constraints that may be very different from training data constraints. For example, the resolution during annotation may differ from the resolution the model trained on. The main tool here is the metadata. As long as it's clear what resolution or accuracy was present when the BLOB was supervised, the rest must be handled on a case by case basis.

Historically, it was common for models to have fixed limits on resolution (accuracy); however, this continues to evolve. From the training data perspective, the main responsibility is to ensure the ML program or ML team knows the level of accuracy at the time of human supervision, and if there are any known assumptions around that accuracy. For example, an assumption might surface that a certain attribute requires a minimum level of resolution to be resolvable by a human.

Some BLOBs introduce extra challenges. A 3D point cloud may contain less data than an image.[10] However, 3D annotation is often more difficult for an end user than

---

10 A point cloud stores triplets for coordinates (x, y, z); there are often a few million of these points. In contrast, a single 4K image contains over eight million RGB (Red Green Blue) triplets.

image annotation. So, while the technical data may be less, the human supervision part may be harder.

Multiple BLOBs may be combined into a compound file. For example, multiple images could be related to each other as a single document. Or an image may be associated with a separate PDF. Conversely, a single BLOB may be split into multiple samples. For example, a large scan might be split into multiple parts.

BLOBs may be transformed between different views. For example, a set of 2D images might be projected into a single 3D frame. Further, this 3D frame may then be annotated from a 2D view. And as a related concept, data may be labeled in a different dimensional space than what it gets trained on. For example, the labels may be done in a 2D context, but trained in 3D space. The main thing to be aware of here is differentiating between what is human supervision and what is a calculated or projected value.

To summarize, there is a high level of complexity to the potential raw data (BLOB) formats and associated artifacts. The main responsibility of training data is to record processing steps done, record relevant assumptions present at human supervision time, and work with data science to ensure alignment between the accuracy available with human supervision and usage in the ML program. By doing this, we recognize the limits of the assumptions we can make and provide everything needed to make the best use of the data and ensure overall system success.

Summary

The schema is one of the most important concepts in training data to understand. Schema is what your ideal machine learning system will achieve. Schema is your meaning of What, Where, and a series of relationships. It's the bridge between raw data, machine learning, and useful predictions.

In this chapter, you learned about important concepts like labels and attributes. We delved into attribute specifics like scope, type, and constraints. You also learned about spatial types like boxes and polygons, and their relation to labels and attributes. We went over computer vision spatial types and concepts around relationships and sequences.

Schema concepts must then be useful to machine learning. I covered common example tasks to illustrate examples of how abstract schema and training data products can be consumed by machine learning. We wrapped up by zooming out to more general concepts, including conceptual boundaries between the production of AI data and consumption by ML systems. Next, we'll take a deep dive into data engineering. We'll cover raw data storage concerns, formatting and mapping, data access, security, and pre-labeling (existing predictions) insertions.

CHAPTER 4

Data Engineering

Introduction

In earlier chapters, you were introduced to abstract concepts. Now, we'll move forward from that technical introduction to discuss implementation details and more subjective choices. I'll show you how we work with the art of training data in practice as we walk through scaling to larger projects and optimizing performance.

Data ingestion is the first and one of the most important steps. And the first step to ingestion is setting up and using a training data *system of record* (SoR). An example of an SoR is a training data database.

Why is data ingestion hard? Many reasons. For example, training data is a relatively new concept, there are a variety of formatting and communication challenges. The volume, variety, and velocity of data vary, and there is a lack of well-established norms, leading to many ways to do it.

Also, there are many concepts, like using a training data database, and who wants to access what when; that may not be obvious, even to experienced engineers. Ingestion decisions ultimately determine query, access, and export considerations.

This chapter is organized into:

- Who wants to use the data and when they want to use it
- Why the data formats and communication methods matter; think "game of telephone"
- An introduction to a training data database as your system of record
- The technical basics of getting started
- Storage, media-specific needs, and versioning

- Commercial concerns of formatting and mapping data
- Data access, security, and pre-labeled data

To achieve a data-driven or data-centric approach, tooling, iteration, and data are needed. The more iteration and the more data, the more need there is for great organization to handle it.

You may ingest data, explore it, and annotate it in that order. Or perhaps you may go straight from ingesting to debugging a model. After streaming to training, you may ingest new predictions, then debug those, then use annotation workflow. The more you can lean on your database to do the heavy lifting, the less you have to do yourself.

Who Wants the Data?

Before we dive into the challenges and the technical specifics, let's set the table about goals and the humans involved here and discuss how data engineering services those end users and systems. After, I'll cover the conceptual reasons for wanting a training data database. I'll frame the need for this by showing what the default case looks like without it, and then what it looks like with it.

For ease of discussion, we can divide this into groups:

- Annotators
- Data scientists
- ML programs (machine to machine)
- Application engineers
- Other stakeholders

Annotators

Annotators need to be served the right data at the right time with the right permissions. Often, this is done at a single-file level, and is driven by very specifically scoped requests. There is an emphasis on permissions and authorization. In addition, the data needs to be delivered at the proper time—but what is the "right time"? Well, in general it means on-demand or online access. This is where the file is identified by a software process, such as a task system, and served with fast response times.

Data scientists

Data science most often looks at data at the set level. More emphasis is placed on query capabilities, the ability to handle large volumes of data, and formatting. Ideally, there is also the ability to drill down into specific samples and compare the results of different approaches both quantitatively and qualitatively.

ML programs

ML programs follow a path similar to that of data science. Differences include permissions schemes (usually programs have more access than individual data scientists), and clarity on what gets surfaced and when (usually more integration and process oriented versus on-demand analysis). Often, ML programs can have a software-defined integration or automation.

Application engineers

Application engineers are concerned with getting data from the application to the training data database and how to embed the annotation and supervision to end users. Queries per second (throughput) and volume of data are often top concerns. There is sometimes a faulty assumption that there is a linear flow of data from an "ingestion" team, or the application, to the data scientists.

Other stakeholders

Other stakeholders with an interest in the training data might be security personnel, DevMLOps professionals, backup systems engineers, etc. These groups often have cross-domain concerns and crosscut other users' and systems' needs. For example, security cares about end-user permissions already mentioned. Security also cares about a single data scientist not being a single point of critical failure, e.g., having the entire dataset on their machine or overly broad access to remote sets.

Now that you have an overview of the groups involved, how do they talk with each other? How do they work together?

A Game of Telephone

Telephone is "a game where you come up with a phrase and then you whisper it into the ear of the person sitting next to you. Next, this person has to whisper what he or she heard in the next person's ear. This continues in the circle until the last person has heard the phrase. Errors typically accumulate in the retellings, so the statement announced by the last player differs significantly from that of the first player, usually with amusing or humorous effect."[1]

As an analogy, you can think of suboptimal data engineering as a game of telephone as shown in Figure 4-1. At each stage, accumulated data errors increase. And it's actually even worse than that graphic shows. This is because sensors, humans, models, data, and ML systems all interact with each other in nonlinear ways.

---

1 From Google Answers

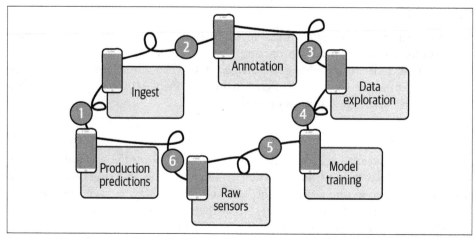

Figure 4-1. Without a system of record, data errors accumulate like a game of telephone

Unlike the game, with training data, these errors are not humorous. Data errors cause poor performance, system degradation, and failures, resulting in real-world physical and financial harm. At each stage, things like different formats, poor or missing data definitions, and assumptions will cause the data to get malformed and garbled. While one tool may know about "xyz" properties, the next tool might not. And it repeats. The next tool won't export all of the properties, or will modify them randomly, etc. No matter how trivial these issues seem on the whiteboard, they will always be a problem in the real world.

Problems are especially prevalent in larger, multiteam contexts and when the holistic needs of all major groups are not taken into account. As a new area with emerging standards, even seemingly simple things are poorly defined. In other words, data engineering is especially important for training data. If you have a greenfield (new) project, then now is the perfect time to plan your data engineering.

How do you know when a system of record is needed? Let's explore that next.

When a system of record is needed

From planning a new project to rethinking existing projects, signals it's time for a training data system of record include the following:

- There is data loss between teams, e.g., because each team owns their own copy of the data.

- Teams aggregate other teams' data but only use a small slice of it, e.g., when a query would be better.

- There is abuse or overuse of unstructured or semi-structured formats like CSVs, strings, JSON, etc., e.g., dumping output into many CSV files in a bucket.

- Cases emerge where the format is only known to the unique application that generated the data. For example, people might be loading unstructured or badly structured data after the fact, hoping for the best.
- There are over assumptions that each system will run in a specific, predefined daisy-chained order instead of a more composable design.
- The overall system performance is not meeting expectations or models are slow to ship or update.

Alternatively, you may simply have no true system of record anywhere. If you have a system, but it does not holistically represent the state of the training data (e.g., thinking of it as just labeling rather than as the center of gravity of the process) then its usefulness is likely low. This is because there will likely be an unreasonable level of communication required to make changes (e.g., changing the schema is not one quick API call or one UI interaction). It also means changes that should be done by end users must be discussed as an engineering-level change.

If each team owns its own copy of the data, there will be unnecessary communications and integration overhead, and likely data loss. This copying is often the "original sin," since the moment there are multiple teams doing this, it will take an engineering-level change to update the overall system—meaning updates will not be fluid, leading to the overall system performance not meeting expectations.

The user expectations and data formats change frequently enough that the solution can't be an overly rigid, automatic process. So don't think about this in terms of automation, but rather in terms of "where is the center of gravity of training data?" It should be with the humans and a system of record, e.g., a training data database, in order to get the best results.

Planning a Great System

So how do you avoid the game of telephone? It starts with planning. Here are a couple of thought starters, and then I'll walk through best practices for the actual setup.

The first is to establish a meaningful unit of work relevant to your business. For example, for a company doing analytics on medical videos, it could be a single medical procedure. Then, within each procedure, think about how many models are needed (don't assume one!), how often they will be updated, how the data will flow, etc. We will cover this in more depth in "Overall system design" on page 176, but for now I just want to make sure it's clear that ingestion is often not a "once and done" thing. It's something that will need ongoing maintenance, and likely expansion over time.

Second, is to think about the data storage and access, setting up a system of record like a training data database. While it is possible to "roll your own," it's difficult to holistically consider the needs of all of the groups. The more a training data database is used, the easier it is to manage the complexity. The more independent storage is used, the more pressure is put on your team to "reinvent the wheel" of the database.

There are some specifics to building a great ingestion subsystem. Usually, the ideal is that these sensors feed directly into a training data system. Think about how much distance, or hops, are there between sensors, predictions, raw data, and your training data tools.

Production data often needs to be reviewed by humans, analyzed at a set level, and potentially further "mined" for improvements. The more predictions, the more opportunity for further system correction. You'll need to consider questions like: How will production data get to the training data system in a useful way? How many times is data duplicated during the tooling processes?

What are our assumptions around the distinctions between various uses of the data? For example, querying the data within a training data tool scales better than expecting a data scientist to export all the data and then query it themselves after.

Naive and Training Data–Centric Approaches

There are two main approaches that people tend to take to working with training data. One I'll refer to as "naive," and the other is more centered on the importance of the data itself (data-centric).

Naive approaches tend to see training data as just one step to be bolted alongside a series of existing steps. Data-centric approaches see the human actions of supervising the data as the "center of gravity" of the system. Many of the approaches in this book align well or equate directly to being data-centric, in some ways making the training-data-first mindset synonymous with data-centric.

For example, a training data database has the definitions, and/or literal storage, of the raw data, annotations, schema, and mapping for machine-to-machine access.

There is naturally some overlap in approaches. In general, the greater the competency of the naive approach, the more it starts to look like a re-creation of a training data–centric one. While it's possible to achieve desirable results using other approaches, it is much easier to consistently achieve desirable results with a training data–centric approach.

Let's get started by looking at how naive approaches typically work.

Naive approaches

Typically, in a naive approach, sensors capture, store, and query the data *independently* of the training data tooling, as shown in Figure 4-2. This usually ends up looking like a linear process, with pre-established start and end conditions.

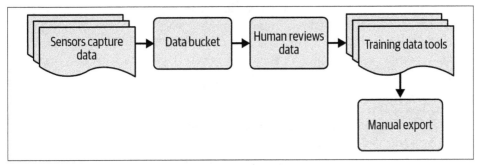

Figure 4-2. Naive data engineering process example

The most common reasons naive approaches get used:

- The project started before training data–centric approaches had mature tooling support.
- Engineers did not know about training data–centric approaches.
- Testing and development of new systems.
- Old, historical data, with no chance of new data coming in (rare).
- Cases where it's impractical to use a training data–centric approach (rare).

Naive approaches tend to look like the game of telephone mentioned earlier. Since each team has its own copy of the data, errors accumulate as that data gets passed around. Since there is either no system of record, or the system of record does not contain the complete state of training data, it is difficult to make changes at the user level. Overall, the harder it is to safely make changes, the slower it is to ship and iterate, and the worse the overall results are.

Additionally, naive approaches tend to be coupled to hidden or undefined human processes. For example, some engineer somewhere has a script on their local machine that does some critical part of the overall flow, but that script is not documented or not accessible to others in a reasonable way. This often occurs because of a lack of understanding of how to use a training data database, as opposed to being a purposeful action.

In naive approaches, there is a greater chance of data being unnecessarily duplicated. This increases hardware costs, such as storage and network bandwidth, in addition to the already-mentioned conceptual bottlenecks between teams. It also can introduce security issues since the various copies may maintain different security postures. For example, a team aggregating or correlating data might bypass security in a system earlier in the processing chain.

A major assumption in naive approaches is that a human administrator is manually reviewing the data (usually only at the set level), so that only the data that appears to be desired to be annotated is imported. In other words, only data pre-designated (usually through a relatively inconsistent means) for annotation is imported. This "random admin before import" assumption makes it hard to effectively supervise production data and use exploration methods, and generally bottlenecks the process because of the manual and undefined nature of the hidden curation process. Essentially this is relying on implicit admin-driven assumptions instead of explicitly defining processes with multiple stakeholders including SMEs. To be clear, it's not the reviewing of the data that's at issue—it's the distinction between a process driven at a larger team level being better than a single admin person working in a more arbitrary manner.

Please think of this conceptually, not in terms of literal automation. A software-defined ingestion process is, by itself, little indication of a system's overall health, since it does not speak to any of the real architectural concerns around the usage of a training data database.

Training data–centric (system of record)

Another option is to use a training data–centric approach. A training data database, as shown in Figure 4-3, is the heart of a training data–centric approach. It forms the system of record for your applications.

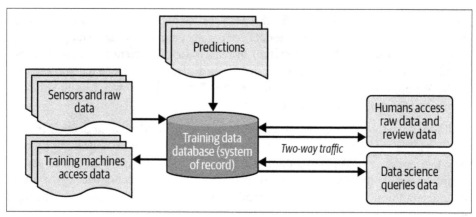

Figure 4-3. Training data database (system of record)

A training data database has the definitions, and/or literal storage, of the raw data, annotations, schema, and mapping for machine-to-machine access, and more. Ideally, it's the complete definition of the system, meaning that given the training data database, you could reproduce the entire ML process with no additional work.

When you use a training data database as your system of record, you now have a central place for all teams to store and access training data. The greater the usage of the database, the better the overall results—similar to how the proper use of a database in a traditional application is well-known to be essential.

The most common reasons to use a training data database:

- It supports a shift to data-centric machine learning. This means focusing on improving overall performance by improving the data instead of just improving the model algorithm.
- It supports more easily aligning multiple ML programs by having the training data definitions all in one place.
- It supports end users supervising (annotating) data and supports embedding of end-user supervision deeper in workflows and applications.
- Data access is now query-based instead of requiring bulk duplication and extra aggregation steps.

Continuing the theme of the distinction between training data and data science, at the integration points, naturally, the line will be the most blurred. In practice, it is reasonable to invoke ML programs from a training data system.

Other reasons for using a training data database include the following:

- It decouples visual UI requirements from data modeling (e.g., query mechanisms).
- It enables faster access to new tools for data discovery, what to label, and more.
- It enables user-defined file types, e.g., representing an "Interaction" as a set of images and text, which supports fluid iteration and end user–driven changes.
- It avoids data duplication and stores external mapping definitions and relationships in one place.
- It unblocks teams to work as fast as they can instead of waiting for discrete stages to be completed.

There are a few problems with a training data database approach:

- Using one requires knowledge of its existence and conceptual understanding.
- Staff need to have the time, ability, and resources to use a training data database.
- Well-established data access patterns may require reworking to the new context.
- Its reliability as a system of record does not have the same history as older systems have.

Ideally, instead of deciding what raw data to send (e.g., from an application to the database), all relevant data gets sent to the database first, before humans select samples to label. This helps to ensure it is the true system of record. For example, there may be a "what to label" program that uses all of the data, even if the humans only review a sample of it. Having all the data available to the "what to label" program through the training data database makes this easy. The easiest way to remember this is to think of the training data database as the center of gravity of the process.

The training data database takes on the role of managing references to, or even contains the literal byte storage of, raw media. This means sending the data directly to the training data tool. In actual implementation, there could be a series of processing steps, but the idea is that the final resting place of the data, the source of truth, is inside the training data database.

A training data database is the complete definition of the system, meaning that given the training data database, you can create, manage, and reproduce all end-user ML application needs, including ML processes, with no additional work. This goes beyond MLOps, which is an approach that often focuses more on automation, modeling, and reproducibility, purely on the ML side. You can think of MLOps as a sub-concern of a more strategic training data–centric approach.

A training data database considers multiple users from day one and plans accordingly. For example, an application designed to support data exploration can establish that indexes for data discovery are automatically created at ingest. When it saves annotations, it can also create indexes for discovery, streaming, security, etc., all at the same time.

A closely related theme is that of exporting the data to other tools. Perhaps you want to run some process to explore the data, and then need to send it to another tool for a security process, such as to blur personally identifiable info. Then you need to send it on to some other firm to annotate, then get the results back from them and into your models, etc. And let's say that at each of these steps, there is a mapping (definitions) problem. Tool A outputs in a different format than Tool B inputs. This type of data transfer often requires orders of magnitude more computing resources than when using other common systems. In a sense, each transfer is more like a mini-database migration, because all the components of the data have to go through the conversion process. This is discussed in "Scale" on page 44 as well.

Generally speaking, the tighter the connection between the sensors and the training data tools, the more potential for all of the end users and tools to be effective together. Every other step that's added between sensors and the tools is virtually guaranteed to be a bottleneck. The data can still be backed up to some other service at the same time, but generally, this means organizing the data in the training data tooling from day one.

The first steps

Let's say you are on board to use a training data–centric approach. How do you actually get started?

The first steps are to:

1. Set up training data database definitions
2. Set up data ingestion

Let's consider definitions first. A training data database puts all the data in one place, including mapping definitions to other systems. This means that there is one single place for the system of record and the associated mapping definitions to modules running within the training data system and external to it. This reduces mapping errors, data transfer needs, and data duplication.

Before we start actually ingesting data, here are a few more terms that need to be covered first:

- Raw data storage
- Raw media BLOB-specific concerns
- Formatting and mapping
- Data access
- Security concerns

Let's start with raw data storage.

Raw Data Storage

The objective is to get the raw data, such as images, video, and text, into a usable form for training data work. Depending on the type of media, this may be easier or harder. With small amounts of text data, the task is relatively easy; with large amounts of video or even more specialized data like genomics, it becomes a central challenge.

It is common to store raw data in a bucket abstraction. This can be on the cloud or using software like MinIO. Some people like to think of these cloud buckets as "dump it and forget it," but there are actually a lot of performance tuning options available.

At the training data scale, raw storage choices matter. There are a few important considerations to keep in mind when identifying your storage solution:

Storage class

There are more major differences between storage tiers than it may first appear. The tradeoffs involve things like access time, redundancy, geo-availability, etc. There are orders-of-magnitude price differences between the tiers. The most key tool to be aware of is lifecycle rules, e.g., Amazon S3's, wherein, usually with a few clicks, you can set policies to automatically move old files to cheaper storage options as they age. Examples of best practices in more granular detail can be found on Diffgram's site (*https://oreil.ly/y8zzM*).

Geolocation (aka Zone, Region)

Are you storing data on one side of the Atlantic Ocean and having annotators access it on the other? It's worth considering where the actual annotation is expected to happen, and if there are options to store the data closer to it.

Vendor support

Not all annotation tools have the same degree of support for all major vendors. Keep in mind that you can typically manually integrate any of these offerings, but this requires more effort than tools that have native integration.

Support for accessing data from these storage providers is different from the tool running on that provider. Some tools may support access from all three, but as a service, the tool itself runs on a single cloud. If you have a system you install on your own cloud, usually the tool will support all three.

For example, you may choose to install the system on Azure. You may then pull data into the tool from Azure, which leads to better performance. However, that doesn't prevent you from pulling data from Amazon and Google as needed.

By Reference or by Value

Storing by reference means storing only small pieces of information, such as strings or IDs, that allow you to identify and access the raw bytes on an existing source system. By value means copying the literal bytes. Once they are copied to the destination system, there is no dependency on the source system.

If you want to maintain your folder structure, some tools support referencing the files instead of actually transferring them. The benefit of this is less data transfer. A downside is that now it's possible to have broken links. Also, separation of concerns could be an issue; for example, some other process may modify a file that the annotation tool expects to be able to access.

Even when you use a pass-by-reference approach for the raw data, it's crucial that the system of truth is the training data database. For example, data may be organized into sets in the database that are not represented in the bucket organization. In addition, the bucket represents only the raw data, whereas the database will have the annotations and other associated definitions.

For simplicity, it's best to think of the training data database as one abstraction, even if the raw data is stored outside the physical hardware of the database, as shown in Figure 4-4.

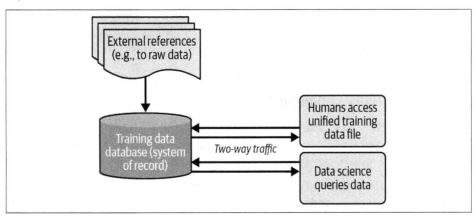

Figure 4-4. Training data database with references to raw data stored externally

Off-the-Shelf Dedicated Training Data Tooling on Your Own Hardware

Let's assume that your tooling is trustworthy (perhaps you were able to inspect the source code). In this context, we trust the training data tool to manage the signed URL process, and to handle the identity and access management (IAM) concerns. With this established, the only real concern is what bucket it uses—and that generally becomes a one-time concern, because the tool manages the IAM. For advanced cases, the tool can still link up with a single sign-on (SSO) or more complex IAM scheme.

Note: The tooling doesn't have to run on the hardware. The next level up is to trust the training data tooling, and also trust the service provider to host/process the data. Although this is an option, keep in mind that it provides the least level of control.

Data Storage: Where Does the Data Rest?

Generally speaking, any form of tooling will generate some form of data that is added to the "original" data. The BLOB data is usually stored by reference, or by physically moving the data.

This means if I have data in bucket A, and I use a tool to process it, there has to either be additional data in bucket A, or I will need a new bucket, B, to be used by the tool. This is true for Diffgram, SageMaker, and, as far as I'm aware, most major tools.

Depending on your cost and performance goals, this may be a critical issue or of little concern. On a practical level, for most use cases, you need to keep in mind only two simple concepts:

- Expect that additional data will be generated.
- Know where the data is being stored, but don't overthink it.

In the same way, we don't really question how much storage, say, a PostgreSQL Write-Ahead Log (WAL) generates. My personal opinion is that it's best to trust the training data tool in this regard. If there's an issue, address it within the training data tool's realm of influence.

External Reference Connection

A useful abstraction to create is a connection from the training data tooling to an existing external reference, e.g., a cloud bucket. There are various opinions on how to do this, and the specifics vary by hardware and cloud provider.

For the non-technical user, this is essentially "logging in" to a bucket. In other words, I create a new bucket, A, and it gets a user ID and password (client ID/client secret) from the IAM system. I pass those credentials to the training data tool and it stores them securely. Then, when needed, the training data tool uses those credentials to interact with the bucket.

Raw Media (BLOB)–Type Specific

A BLOB is a Binary Large Object. It is the raw data form of media, and is also known as an "object." The technology that stores raw-data BLOBs is called a "bucket" or an "object store." Depending on your media type, there will be specific concerns. Each media type will have vastly more concerns than can be listed here. You must research each type for your needs. A training data database will help format BLOBs to be useful for multiple end users, such as annotators and data scientists. Here, I note some of the most common considerations to be aware of.

Images

Images usually don't require any technical data prep because they are individually small enough files. More complex formats like TIF will often be "flattened," although it's possible to preserve layers in some tools.

Video

It's common to split video files into smaller clips for easier annotation and processing. For example, a 10-minute video might be split into 60-second clips.

Sampling frames is one approach to reduce processing overhead. This can be done by reducing the frame rate and extracting frames. For example, you can convert a file from 30 frames per second (FPS) to 5 or 10 per second. A drawback is that losing too many frames may make it more difficult to annotate (for example, a key moment in the video may be cut off) or you may lose the ability to use other relevant video features like interpolation or tracking (that rely on assumptions around having a certain frame rate). Usually, it's best to keep the video as a playable video and also extract all the frames required. This improves the end-user annotation experience and maximizes annotation ability.

Event-focused analytics need the exact frame when something happens, which effectively becomes lost if many frames are removed. Further, with all the frames intact, the full data is available to be sampled by "find interesting highlights" algorithms. This leads annotators to see more "interesting" things happening, and thus, higher quality data. Object tracking and interpolation further drive this point home, as an annotator may only need to label a handful of frames and can often get many back "for free" through those algorithms. And while, in practice, nearby frames are generally similar, it often still helps to have the extra data.

An exception to this is that sometimes, a very high FPS video (e.g., 240–480+) may still need to be sampled down to 120 FPS or similar. Note that just because many frames are available to be annotated, we can still choose to train models only on completed videos, completed frames, etc. If you must downsample the frames, use the global reference frame to maintain the mapping of the downsampled frame to the original frame.

3D

Usually, you will need to transmit each file in a series of x,y,z triples (3D points) to the SDK.

Text

You will need to select your desired tokenizer or confirm that the tokenizer used by the system meets your needs. The tokenizer divides up words, for example based on spaces or using more complex algorithms, into smaller components. There are many open source tokenizers and a more detailed description of this is beyond the scope of this book. You may also need a process to convert BLOB files (e.g., .txt) into strings or vice versa.

Medical

If your specific medical file is not supported by the tool, you may need to downsample color channels, select which z-axis or slice you wish to use, and crop out images from a too-large single image.

Geospatial

GeoTiff and Cloud-Optimized GeoTIFF (COG) are standard formats in geospatial analysis. Not all training data tools support either format; however, when support is offered, the trend appears to be for COG. Be aware that a projection mapping change may be needed to standardize layers.

Formatting and Mapping

Raw media is one part of the puzzle. Annotations and predictions are another big part. Think in terms of setting up data definitions rather than a one-time import. The better your definitions, the more easily data can flow between ML applications, and the more end users can update the data without engineering.

User-Defined Types (Compound Files)

Real-world cases often involve more than one file. For example, a driver's license has a front and a back. We can think of creating a new user-defined type of "driver's license" and then having it support two child files, each being an image. Or we can consider a "rich text" conversation that has multiple text files, images, etc.

Defining DataMaps

A DataMap handles the loading and unloading of definitions between applications. For example, it can load data to a model training system or a "What to Label" analyzer. Having these definitions well defined allows smooth integrations by end users, and decouples the need for engineering-level changes. In other words, it decouples when an application is called from the data definition itself. Examples include declaring a mapping between spatial locations formatted as x_min, y_min, x_max, y_max and top_left, bottom_right, or mapping integer results from a model back to a schema.

Ingest Wizards

One of the biggest gaps in tooling is often around the question, "How hard is it to set up my data in the system and maintain it?" Then comes what type of media can it ingest? How quickly can it ingest it?

This is a problem that's still not as well defined as in other forms of software. You know when you get a link to a document and you load it for the first time? Or some big document starts to load on your computer?

Recently, new technology like "Import Wizards"—step-by-step forms—have come up that help make some of the data import process easier. While I fully expect these processes to continue to become even easier over time, the more you know about the

---

behind-the-scenes aspects, the more you understand how these new wonderful wizards are actually working. This started originally with file browsers for cloud-based systems, and has progressed into full-grown mapping engines, similar to smart switch apps for phones, like the one that allows you to move all your data from Android to iPhone, or vice versa.

At a high level how it works is that a mapping engine (e.g., part of the ingest wizard) steps you through the process of mapping each field from one data source to another. Mapping wizards offer tremendous value. They save having to do a more technical integration. They typically provide more validations and checks to ensure the data is what you expect (picture seeing an email preview in Gmail before committing to open the email). And best of all, once the mappings are set up, they can easily be swapped out from a list without any context switching!

The impact of this is hard to overstate. Before, you may have been hesitant to, say, try a new model architecture, commercial prediction service, etc., because of the nuances of getting the data to and from it. This dramatically relieves that pressure.

What are the limitations of wizards? Well, first, some tools don't support them yet, so they simply may not be available. Another issue is that they may impose technical limitations that are not present in more pure API calls or SDK integrations.

Organizing Data and Useful Storage

One of the first challenges is often how to organize the data you have already captured (or will capture). One reason this is more challenging than it may at first appear is that often these raw datasets are stored remotely.

At the time of writing, cloud data storage browsers are generally less mature than local file browsers. So even the most simple operations, e.g., me sitting at a screen and dragging files, can provide a new challenge.

Some practical suggestions here:

Try to get the data into your annotation tool sooner in the process than later. For example, at the same time new data comes in, I can write the data reference to the annotation tool at a similar time at which I'm writing to a generic object store. This way, I can "automatically" organize it to a degree, and/or more smoothly enlist team members to help with organization-level tasks.

Consider using tools that help surface the "most interesting" data. This is an emerging area—but it's already clear that these methods, while not without their challenges, have merit and appear to be getting better.

Use tags. As simple as it sounds, tagging datasets with business-level organizational information helps. For example, the dataset "Train Sensor 12" can be tagged

"Client ABC." Tags can crosscut data science concerns and allow both business control/organizational and data science-level objectives.

Remote Storage

Data is usually stored remotely relative to the end user. This is because of the size of the data, security requirements, and automation requirements (e.g., connecting from an integrated program, practicalities of running model inference, aggregating data from nodes/system). Regarding working in teams, the person who administers the training data may not be the person who collected the data (consider use cases in medical, military, on-site construction, etc.).

This is relevant even for solutions with no external internet connection, also commonly referred to as "air-gapped" secret-level solutions. In these scenarios, it's still likely the physical system that houses the data will be in a different location than the end user, even if they are sitting two feet from each other.

The implication of the data being elsewhere is that we now need a way to access it. At the very least, whoever is annotating the data needs access, and most likely some kind of data prep process will also require it.

Versioning

Versioning is important for reproducibility. That said, sometimes versioning gets a little too much attention. In practice, for most use cases, being mindful of the high-level concepts, using snapshots, and having good overall system of record definitions will get you very far.

There are three primary levels of data versioning, per instance (annotation), per file, and export. Their relation to each other is shown in Figure 4-5.

Figure 4-5. Versioning high-level comparison

Next, I'll introduce each of these at a high level.

Per-instance history

By default, instances are not hard deleted. When an edit is made to an existing instance, Diffgram marks it as a soft delete and creates a new instance that succeeds it, as shown in Figure 4-6. You could, for example, use this for deep-dive annotation or model auditing. It is assumed that `soft_deleted` instances are not returned when the default instance list is pulled for a file.

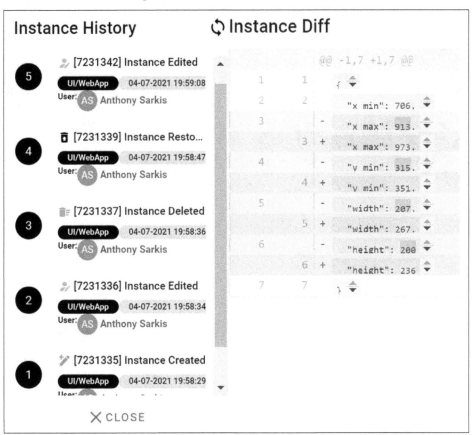

Figure 4-6. Left, per-instance history in UI; right, a single differential comparison between the same instance at different points in time

Per file and per set

Each set of tasks may be set to automatically create copies per file at each stage of the processing pipeline. This automatically maintains multiple file-level versions relevant to the task schema.

You may also programmatically and manually organize and copy data into sets on demand. You can filter data by tags, such as by a specific machine learning run. Then, compare across files and sets to see what's changed.

Add files to multiple sets for cases where you want the files to always be on the latest version. That means you can construct multiple sets, with different criteria, and instantly have the latest version as annotation happens. Crucially, this is a living version, so it's easy to always be on the "latest" one.

You can use these building blocks to flexibly manage versions across work in progress at the administrator level.

Per-export snapshots

With per-export snapshots, every export is automatically cached into a static file. This means you can take a snapshot at any moment, for any query, and have a repeatable way to access that exact set of data. This can be combined with webhooks, SDKs, or userscripts to automatically generate exports. These can be generated on demand, anytime. For example, you can use per-export snapshots to guarantee that a model is accessing the exact same data. The header of an export is shown as an example in Figure 4-7.

Figure 4-7. Export UI list view example

Next, we'll cover exporting and accessing pattern trade-offs in more detail in Data Access.

Data Access

So far, we have covered overall architectural concepts, such as using a training data database to avoid a game of telephone. We covered the basics of getting started, media storage, mapping concepts, and BLOBs and other formats. Now we discuss the annotations themselves. A benefit of using a training data–centric approach is you get best practices, such as snapshots and streaming, built-in.

Streaming is a separate concept from querying, but it goes hand in hand in practical applications. For example, you may run a query that results in 1/100 of the data of a file-level export, and then stream that slice of the data directly in code.

There are some major concepts to be aware of:

- File-based exports
- Streaming
- Querying data

Disambiguating Storage, Ingestion, Export, and Access

One way to think of this is that often, the way data is utilized in a database is different from how it's stored at rest, and how it's queried. Training data is similar. There are processes to ingest data, different processes to store it, and different ones again to query it:

- Raw data storage refers to the literal storage of the BLOBs, and not the annotations, which are assumed to be kept in a separate database.

- Ingestion is about the throughput, architecture, formats, and mapping of data. Often, this is between other applications and the training data system.

- Export, in this context, usually refers to a one-time file-based export from the training data system.

- Data access is about querying, viewing, and downloading BLOBs and annotations.

Modern training data systems store annotations in a database (not a JSON dump), and provide abstract query capabilities on those annotations.

File-Based Exports

As mentioned in versioning, a file-based export is a moment-in-time snapshot of the data. This is usually generated only on a very rough set of criteria, e.g., a dataset name. File-based exports are fairly straightforward, so I won't spend much time on them. A comparison of trade-offs of file-based exporting and streaming is covered in the next section.

Streaming Data

Classically, annotations were always exported into a static file, such as a JSON file. Now, instead of generating each export as a one-off thing, you stream the data directly into memory. This brings up the question, "How to get my data out of the system?" All systems offer some form of export, but you'll need to consider what kind. Is it a static, one-term export? Is it direct to TensorFlow or PyTorch memory?

Streaming benefits

- You can load only what you need, which may be a small percent of a JSON file. At scale, it may be impractical to load all the data into a JSON file, so this may be a major benefit.

- It works better for large teams. It avoids having to wait for static files; you can program and do work with the expected dataset before annotation starts, or even while annotation is happening.

- It's more memory efficient—because it's streaming, you need not ever load the entire dataset into memory. This is especially applicable for distributed training, and when marshaling a JSON file would be impractical on a local machine.
- It saves having "double mapping," e.g., mapping to another format, which will itself then be mapped to tensors. In some cases too, parsing a JSON file can take even more effort than just updating a few tensors.
- It provides more flexibility; the format can be defined and redefined by end users.

Streaming drawbacks

- The specifications are defined in the code. If the dataset changes, reproducibility may be affected unless additional steps are taken.
- It requires a network connection.
- Some legacy training systems/AutoML providers may not support loading directly from memory and may require static files.

One thing to keep top of mind throughout this is that we don't really want to statically select folders and files. We are really setting up a process in which we stream new data, in an event-driven way. To do this, we need to think of it more like assembling a pipeline, rather than focusing on the mechanics of getting a single existing known set.

Example: Fetch and stream

In this example, we will fetch a dataset and stream it using the Diffgram SDK:

```
pip install diffgram==0.15.0

from diffgram import Project
project = Project(project_string_id='your_project')
default_dataset = project.directory.get(name = 'Default')
# Let's see how many images we got
print('Number of items in dataset: {}'.format(len(default_dataset)))
# Let's stream just the 8th element of the dataset
print('8th element: {}'.format(default_dataset[7]))
pytorch_ready_dataset = default_dataset.to_pytorch()
```

Queries Introduction

Each application has its own query language. This language often has a special structure specific to the context of training data. It also may have support for abstract integrations with other query constructs.

To help frame this, let's start with this easy example to get all files with more than 3 cars and at least 1 pedestrian (assuming those labels exist in your project):

```
dataset = project.dataset.get('my dataset')
sliced_dataset = dataset.slice('labels.cars > 3 and labels.pedestrian >= 1')
```

Integrations with the Ecosystem

There are many applications available to perform model training and operation. At the time of writing, there are many hundreds of tools that fall in this category. As mentioned earlier, you can set up a mapping of definitions for formats, triggers, dataset names, and more in your training data tooling. We will cover these concepts in more depth in later chapters.

Security

The security of training data is of critical importance. Often, raw data is treated with greater scrutiny than data in other forms from a security viewpoint. For example, the raw data of critical infrastructure, driver's licenses, military targets, etc. is very carefully stored and transferred.

Security is a broad topic that must be researched thoroughly and addressed separately from this book. However, we will address some issues that are crucial to understand when working with training data. I am calling attention specifically to the most common security items in the context of data engineering for training data:

- Access control
- Signed URLs
- Personally identifiable information

Access Control

There are a few major questions about access control that we can focus on to start. What system is processing the data? What are the identity and access management (IAM) permission concerns around system-level processing and storage? What user access concerns are there?

Identity and Authorization

Production-level systems will often use OpenID Connect (OIDC). This can be coupled with role-based access control (RBAC) and attribute-based access control (ABAC).

Specifically with regard to training data, often the raw data is where there is the most tension around access. In that context, it can usually be addressed on either the per-file or per-set level. At the per-file level, access must be controlled by a policy engine that is aware of the triplicate of the {user, file, policy}. This can be complex to administer at the per-file level. Usually, it is easier to achieve this at the per-set level. At the set (dataset) level, it is achieved with {user, set, policy}.

Example of Setting Permissions

In this code sample, we will create a new dataset and a new security advisor role, and add the {view, dataset} permission abstract object pair on that role:

```
restricted_ds1 = project.directory.new(name='Hidden Dataset 1',
    access_type='restricted')
advisor_role = project.roles.new(name='security_advisor')
advisor_role.add_permission(perm='dataset_view', object_type='WorkingDir')
```

We then assign the user (member) to the restricted dataset:

```
member_to_grant = project.get_member(email='security_advisor_1@example.com')
advisor_role.assign_to_member_in_object(member_id=member.get('member_id'),
    object_id=restricted_ds1.id, object_type='WorkingDir')
```

Alternatively, this can be done with an external policy engine.

Signed URLs

Signed URLs are a technical mechanism to provide secure access to resources, most often raw media BLOBs. A signed URL is the output of a security process that involves identity and authorization steps. A signed URL is most concretely thought of as a one-time password to a resource, with the most commonly added caveat that it expires after a preset amount of time. This expiry time is sometimes as short as a few seconds, is routinely under one week, and rarely, the signed URL will be applicable "virtually indefinitely," such as for multiple years. Signed URLs are not unique to training data, and you might benefit from further researching them, as they appear to be simple but contain many pitfalls. We touch on signed URLs here only in the context of training data.

One of the most critical things to be aware of is that because signed URLs are ephemeral, it is not a good idea to transmit signed URLs as a one-time thing. Doing so would effectively cripple the training data system when the URL expires. It is also less secure, since either the time will be too short to be useful, or too long to be secure. Instead, it is better to integrate them with your identity and authorization system. This way, signed URLs can be generated on demand by specific {User, Object/Resource} pairs. That specific user can then get a short-expiring URL.

In other words, you can use a service outside the training data system to generate the signed URLs so long as the service is integrated directly with the training data system. Again, it's important to move as much of the actual organizational logic and definitions inside the training data system as possible. Single sign-on (SSO) and identity and access–management integration commonly crosscut databases and applications, so that's a separate consideration.

In addition to what we cover in this section, training data systems are currently offering new ways to secure data. This includes things like transmitting training data directly to ML programs, thus bypassing the need for a single person to have extremely open access. I encourage you to read the latest documentation from your training data system provider to stay up to date on the latest security best practices.

Cloud connections and signed URLs

Whoever is going to supervise the data needs to view it. This is the minimum level of access and is essentially unavoidable. The prep systems, such as the system that removes personally identifiable information (PII), generate thumbnails, pre-labels, etc., also need to see it. Also, for practical system-to-system communication, it's often easier to transmit only a URL/file path and then have the system directly download the data. This is especially true because many end-user systems have much slower upload rates than download rates. For example, imagine saying "use the 48 GB video at this cloud path" (KBs of data) versus trying to transmit 48 GB from your home machine.

There are many ways to achieve this, but signed URLs—a per-resource password system—are currently the most commonly accepted method. They can be "behind the scenes," but generally always end up being used in some form.

For both good and bad reasons, this can sometimes be an area of controversy. I'll highlight some trade-offs here to help you decide what's relevant for your system.

Signed URLs

A signed URL is a URL that contains both the location of a resource (like an image) and an integrated password. It's similar to a "share this link" in Google Docs. Signed URLs may also contain other information and commonly time decay, meaning the password expires. For example, a signed URL might have the general form of: *sample.com/123/?password=secure_password*. Please note: Actual signed URLs are usually quite long, about the length of this paragraph or more.

To sum up, training data presents some unusual data processing security concerns that must be kept in mind:

- Humans see the "raw" data in ways that are uncommon in other systems.
- Admins usually need fairly sweeping permissions to work with data, again in ways uncommon in classic systems.

- Due to the proliferation of new media types, formats, and methods of data transmission, there are size and processing concerns with regard to training data. Although the concerns are similar to ones we're familiar with from the classic systems, there are fewer established norms for what's reasonable.

Personally Identifiable Information

Personally identifiable information must be handled with care when working with training data. Three of the most common ways to address PII concerns are to create a PII-compliant handling chain, to avoid it altogether, or to remove it.

PII-compliant data chain

Although PII introduces complications into our workflows, sometimes its presence is needed. Perhaps the PII is desired or useful for training. This requires having a PII-compliant data chain, PII training for staff, and appropriate tagging identifying the elements that contain PII. This also applies if the dataset contains PII and the PII will not be changed. The main factors to look at here are:

- OAuth or similar Identify methods, like OIDC
- On-premises or cloud-oriented installs
- Passing by reference and not sending data

PII avoidance. It may be possible to avoid handling PII. Perhaps your customers or users can be the ones to actually look at their own PII. This may still require some level of PII compliance work, but less than if you or your team directly looks at the data.

PII removal. You may be able to strip, remove, or aggregate data to avoid PII. For example, the PII may be contained in metadata (such as DICOM). You may wish to either completely wipe this information, or retain only a single ID linking back to a separate database containing the required metadata. Or, the PII may be contained in the data itself, in which case other measures apply. For images, for example, this may involve blurring faces and identifying marks such as house numbers. This will vary dramatically based on your local laws and use case.

Pre-Labeling

Supervision of model predictions is common. It is used to measure system quality, improve the training data (annotation), and alert on errors. I'll discuss the pros and cons of pre-labeling in later chapters; for now, I'll provide a brief introduction to the technical specifics. The big idea with pre-labeling is that we take the output from an already-run model and surface it to other processes, such as human review.

Updating Data

It may seem strange to start with the update case, but since records will often be updated by ML programs, it's good to have a plan for how updates will work prior to running models and ML programs.

If the data is already in the system, then you will need to refer to the file ID, or some other form of identification such as the filename, to match with the existing file. For large volumes of images, frequent updates, video, etc., it's much faster to update an existing known record than to reimport and reprocess the raw data.

It's best to have the definitions between ML programs and training data defined in the training data program.

If that is not possible, then at least include the training data file ID along with the data the model trains on. Doing this will allow you to more easily update the file later with the new results. This ID is more reliable than a filename because filenames are often only unique within a directory.

Pre-labeling gotchas

Some formats, for example video sequences, can be a little difficult to wrap one's head around. This is especially true if you have a complex schema. For these, I suggest making sure the process works with an image, and/or the process works with a single default sequence, before trying true multiple sequences. SDK functions can help with pre-labeling efforts.

Some systems use relative coordinates, and some use absolute ones. It is easy to transform between these, as long as the height and width of the image are known. For example, a transformation from absolute to relative coordinates is defined as "x / image width" and "y / image height." For example: A point x,y (120, 90) with an image width/height (1280, 720) would have a relative value of 120/1280 and 90/720 or (0.09375, 0.125).

If this is the first time you are importing the raw data, then it's possible to attach the existing instances (annotations) at the same time as the raw data. If it's not possible to attach the instances, then treat it as updating.

A common question is: "Should all machine predictions be sent to the training data database?" The answer is yes, as long as it is feasible. Noise is noise. There's no point in sending known noisy predictions. Many prediction methods generate multiple predictions with some threshold for inclusion. Generally, whatever mechanism you have for filtering this data needs to be applied here too. For example, you might only take the highest "confidence" prediction. To this same end, in some cases, it can be very beneficial to include this "confidence" value or other "entropy" values to help better filter training data.

Pre-labeling data prep process

Now that we have covered some of the abstract concepts, let's dive into some specific examples for selected media formats. We cannot cover all possible formats and types in this book, so you must research the docs (*https://oreil.ly/VlQQo*) for your specific training data system, media types, and needs.

Figure 4-8 shows an example pre-labeling process with three steps. It's important to begin by mapping your data to the format of your system of record. Once your data has been processed, you'll want to be sure to verify that everything is accurate.

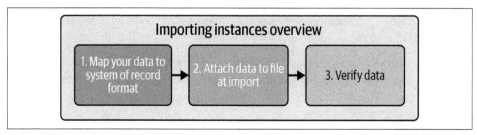

Figure 4-8. Block diagram example

Usually, there will be some high-level formatting information to note, such as saying that an image may have many instances associated with it, or that a video may have many frames, and each frame may have many instances, as shown in Figure 4-9.

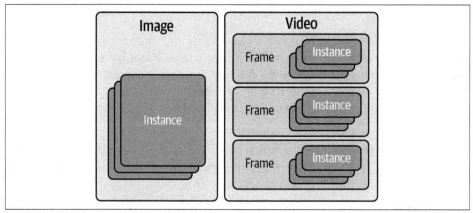

Figure 4-9. Visual overview of relationship between raw media and instances

Let's put all of this together into a practical code example that will mock up data for an image bounding box.

Here is example Python code:

```python
def mock_box(
            sequence_number: int = None,
            name : str = None):

    return {
        "name" : name,
        "number": sequence_number,
        "type": "box",
        "x_max": random.randint(500, 800),
        "x_min": random.randint(400, 499),
        "y_max": random.randint(500, 800),
        "y_min": random.randint(400, 499)
        }
```

This is one "instance." So for example, running the function mock_box() will yield the following:

```python
instance = {
    "name" : "Example",
        "number": 0,
        "type": "box",
        "x_max": 500,
        "x_min": 400,
        "y_max": 500,
        "y_min": 400
}
```

We can combine instances into a list to represent multiple annotations on the same frame:

```python
instance = {}
instance_list = [instance, instance, instance]
```

Summary

Great data engineering for training data requires a central system of record, raw data considerations, ingestion and query setups, defining access methods, securing data appropriately, and setting up pre-labeling (e.g., model prediction) integrations. There are some key points from this chapter that merit review:

- A system of record is crucial to achieve great performance and avoid accumulating data errors like in a game of telephone.
- A training data database is an example of a practical system of record.
- Planning ahead for a training data system of record is ideal, but you can also add it to an existing system.
- Raw data storage considerations include storage class, geolocation, cost, vendor support, and storing by reference versus by value.

- Different raw data media types, like images, video, 3D, text, medical, and geospatial data have particular ingestion needs.

- Queries and streaming provide more flexible data access than file exports.

- Security aspects like access control, signed URLs, and PII must be considered.

- Pre-labeling ideally loads model predictions into the system of record.

- Mapping formats, handling updates, and checking accuracy are key parts of pre-labeling workflows.

Using the best practices in this chapter will help you on your path to improving your overall training data results and the machine learning model consumption of the data.

CHAPTER 5

Workflow

Introduction

Training data is about creating human meaning with data. Humans are, naturally, a vital component of that. In this chapter, I will cover the nuts and bolts of the human workflow of training data.

I will first provide a brief overview of how workflow is the glue between tech and people. I start with motivations for human tasks and move on to the core themes of workflow:

- Getting started
- Quality assurance
- Analytics and data exploration
- Data flow
- Direct annotation

In "Getting Started with Human Tasks" on page 132 I'll talk about the basics, things like why schemas tend to stick around, user roles, training, and more. The next most crucial thing to understand is quality assurance (QA). I focus on the structural level of things, thinking about important motivations for having trust in your human annotators, the standard review loop, and common causes of errors.

After you have started and done some basic QA, you will want to start learning about how to analyze your tasks, datasets and more. This section leads into using models to debug your data, and more generally, how to work with models.

Data flow, getting data moving and in front of humans, and then to models, is a key part of workflow.

Finally, I will wrap up the chapter by taking a deep dive into direct annotation itself. This will cover high-level concepts like business process integration, supervising existing data, and interactive automations, as well as a detailed example of video annotation.

Glue Between Tech and People

Between the data engineering, and the human tasks themselves, is a concept I will refer to here as workflow.

Workflow is all of the definitions, and the "glue," that happens in between technical data connections and the associated human tasks.

For example, data engineering may connect a bucket to your training data platform. But how do you decide when to pull that data into tasks? What do you do after those tasks are complete? Good workflows move data and processes in the right direction before and after human tasks are completed.

The code needed to implement these admin decisions, this glue is often made up of ad hoc notes, one-off scripts, and other fairly brittle artifacts and processes. Further complicating this is a growing selection of intermediary steps, such as running privacy filters, pre-labeling, routing or sorting data, and integrating with third-party business logic.

Instead, a good workflow will generally aim for the following characteristics:

- Clearly defined processes, surfacing as much of the glue code in between stages as possible
- Human tasks are explicitly included
- A well-understood timing protocol: what is manual, what is automated, and everything in between
- A clearly defined export step including what datasets or slices of the data (e.g., data queries) are used
- All third-party steps and integrations, such as webhooks, training systems, pre-labeling, etc., are clearly surfaced
- Has a clear system boundary or "leave off" point, for example when it is connecting to a large orchestration system or model training system
- Flexible enough that admins can make substantial changes to it with minimal IT support (for example, abstracting a data connection from how it is used in a workflow)

You may define some or all of these steps in your training data platform. In that case, there may be built-in options for the timing of the workflow. For example, you could set it up such that each step in a workflow will complete when a substep completes, when a whole step completes, on a predefined schedule, or only when manually triggered.

The specific implementation details of this workflow are, naturally, highly dependent on your specific organization and tooling selection. By the very nature of this glue, it will be different in nearly every case. Given that, the key thing is to simply be aware that such glue and workflow exists, and that aspects of the skeleton of the workflow can be placed directly in the training data system.

In this chapter I will mostly focus on the tasks part of workflow, depicted in Figure 5-1, as that is the most critical and most well defined. I will briefly review some of the other common workflow steps and "glue" that are needed to make this work.

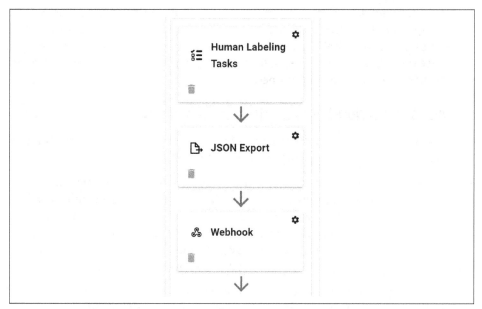

Figure 5-1. Example of a workflow

There are a few key takeaways about considering workflow in your system:

- You can think of workflow as the glue between everything else in training data.
- Within a workflow, most of the focus of attention is on the human-task-related concepts.
- The use of substeps is one the principles of workflow. Technical integrations, automations, and data connections, are some examples.

- A workflow, even if it's just a collection of manual scripts and one-off manual steps by specific people, is always present, regardless of how much is clearly defined in a single system.

- A training data system can provide the architectural structure of a workflow, but each substep is different and specific to your cases.

Note that the glue can also be between tech and tech.

Having this glue code well defined in the training data system is a relatively new area. I won't cover it extensively in this book given how quickly it is changing; please refer to the documentation of your provider for more information.

Why Are Human Tasks Needed?

There are a couple of ways to think about the importance of human workers in the context of training data. First, there are usually a lot of frontline people who review, mold, shape, and digest the raw data into new forms. This naturally leads to the need for some type of organization. Second, most existing generic task systems simply don't have the level of support needed to work so closely with data and dance between the human and technical definitions needed for AI.

Partnering with Non-Software Users in New Ways

Annotating data is similar to doing any technically demanding task. Administrators face some big challenges organizing and managing this work.

If you are also a subject matter expert, then you may be able to create and update the schema mostly on your own. However, there may be a gap in understanding how to match the schema to a desired data science modeling need, requiring more interfacing with the data science. Conversely, if you are a data scientist, you may need to lean heavily on your subject matter experts to build an accurate schema. Either way, you will be partnering with your annotators to jointly create a "codebase."

Generally, the person who administers a project like this at scale will work closely with all stakeholders, including engineers, subject matter experts, and data scientists.

Getting Started with Human Tasks

Most of the workflow setup can be achieved without a deep technical knowledge; however, that still presumes that there is a system ready and available to be configured. Therefore, I will also be assuming that you already have a training data platform up and running, and that the technical items, such as those covered in Chapter 4, are ready to go. You'll get the most out of this chapter if you review what we covered in Chapters 1 through 4 before diving in.

Basics

To set up your human tasks, there are a few critical steps that must be completed:

- Naming tasks
- Selecting a schema
- Determining who can annotate
- Determining the flow of data
- Launching

There are also many optional steps, such review loops, UI customization, and more.

In this chapter's discussion of workflow, I am focusing on the human tasks themselves. All of these steps should actually be very quick and straightforward, since most of the configuration work should be already done.

The first two steps are to choose a good name for your tasks, and decide who can annotate. Both steps will be dependent on your specific business context. The name is, naturally, used to organize the work, and most systems will support some kind of additional task tagging, for example for cost centers, or other project metadata.

To briefly recap, a schema is a paradigm for encoding Who, What, Where, How, and Why. It's a representation of meaning structured by labels, attributes, and their relations to each other. Choosing the schema can mean creating a new one, or it can mean choosing an already created schema, as we cover in depth in Chapter 3.

It is true that in easy cases, a project manager or admin may edit a schema or create a new one while in the process of creating tasks. However, for most real-world business cases, the schema creation is a complex process. For example, it may be populated by an API/SDK, involve sign-offs by multiple people, and take hours, days, or even weeks to create. Therefore, for the scope of a smaller-scale project, such as an admin creating human tasks, you'll want to select an existing schema, rather than creating your own.

Dataflow refers to the selection of an already defined dataset within the training data program. This could be a dataset that already has data, or an empty set that is expected to receive new data in the future, e.g., through a technical integration. I will cover dataset selection, as it relates to human tasks, more in "Dataflow" on page 148.

The technical integrations of dataflow should already be set up, as covered in Chapter 4. I repeat this theme to underscore the different roles of the people involved. An admin may be selecting a dataset, but the meaning of that dataset should already be well defined in the system by data engineering. Data science will be slicing and viewing the data across multiple sets and human tasks, so ideally they should be only minimally concerned with the dataflow in the tasks stage. Once all the other steps are completed, tasks are launched, making them fully available to annotators.

As mentioned in the workflow intro, human tasks can be thought of as the most major building block in a workflow. Other building blocks, such as automation concepts, are covered in Chapter 8. Zooming out, conceptual-level ideas around AI transformation topics will be covered in Chapter 7.

Schemas' Staying Power

When considering tooling, it's important to consider that what is set up in the beginning of a process is likely to persist. One of the reasons that schemas are so impactful is their tendency to persist in systems. This staying power is due to several factors:

- A large time commitment may be needed to create a substantial initial schema, leading to reluctance to institute a new one.
- Schemas can be changed or expanded nearly anytime, so can accommodate changing circumstances.
- Changes to the schema may invalidate portions of prior work.

To illustrate this, consider a schema with one label: "fruit tree." A "fruit tree" gets annotated on a few samples. The schema is then changed to "*apple* tree." At the time of labeling (or model prediction) the annotator or model was thinking about "any fruit trees," not "*apple* trees."

Leaning into attributes and keeping labels generic usually helps make changes smoother. For example, the more generic top-level label of "tree" instead of "fruit tree" keeps it open, so if in the future we add a type (e.g., "apple," "pear," "non-fruit," etc.) we can smoothly expand the existing annotations. This also works if the attribute is added up front and made optional. If the tree type hasn't been determined yet, it can be null.

Of course, changes may be intentional, but even if we intentionally, say, remove "tree" from our schema, there may be existing training data records to update (or delete), and there will be historical predictions that used that schema. Further, if there's a compliance requirement to keep logs of the predictions (and schemas), then that schema may stick around for years. Zooming back to day-to-day work, the key thing to remember as an admin is that it's important to be aware of a schema's staying power, and that it will likely be changing over time. This awareness can affect how you manage human tasks.

User Roles

When considering system roles, there are generally two major buckets to look at—predefined roles and custom roles.

The most common predefined roles include:

- Super-admin: Installation-wide administrator
- Admin: All access to the project
- Editor: Data engineering and data science access
- Annotator: Annotation access
- Viewer: View-only access

In general, most of your users will be assigned the annotation permission, since they should only need to annotate.

Roles are scoped to a specific project default.

Predefined roles may still be integrated with an SSO system, for example mapping frontline worker access to the annotator role.

Custom roles are more complex. A custom role may enforce permissions through a policy engine or external system. You may also create a custom role within the training data system, and then attach permissions to specific objects. Custom roles can quickly become very complex, and a detailed treatment is outside the scope of this book; refer to the documentation of your training data platform.

Training

For most businesses, you will need to conduct some type of training sessions with frontline staff, especially annotators. One approach is to have this conducted by the admin or manager of the frontline staff.

An example Training Plan, using this book as recommended reading, may look like:

1. Training Data Introduction (10 minutes)

   Reading excerpts or domain-specific summaries from Chapter 1
2. Schema (10 minutes)

   Reading excerpts from Chapter 3
3. Annotation Tooling (15 minutes)

   Demonstrating how to use the specific annotation tooling software

   Hotkeys and attribute hotkeys
4. Business-Specific Expectations (5 minutes)

5. General Q&A (10 minutes)

6. Support (10 minutes)

   How to access the help desk

7. Planning for the Desired Next Session (5 minutes)

Gold Standard Training

Let's start from the assumption that annotators are well-meaning and, in general, will make good choices based on the data presented. In this context, an Administrator is still responsible to provide a definition of what "correct" is. These definitions can be fluid. They can be developed by the most senior subject matter experts. This approach, with definitions that are sometimes referred to as the "Gold Standard," can be combined with random sampling of "review" type tasks. It can also be used as an examination. For example, an image is shown to an annotator, they do their attempt, and then the result is compared to the Gold Standard.

The creation of Gold Standards usually helps surface issues in the schema and data. This is because the subject matter experts and admins who create them must have actual annotations to go along with the schema to show what "correct" looks like. Gold Standard approaches have a more clear "right/wrong."

In comparison to "consensus" approaches (multiple people annotating), the Gold Standard method is usually a lower cost. And yet, generally, it can yield similar or even better results. There is more accountability with a Gold Standard. If I create a task, and it gets reviewed and there's an issue, that's a clear signal. It also helps encourage individual learning experiences—e.g., annotators improving. This is a developing area, and there may be other approaches that come up in the future. This area is also somewhat controversial.

Task Assignment Concepts

The most common way to assign tasks include:

- Automatic, "on demand" by annotators' or applications' requests; default
- Predetermined assignment, e.g., round-robin
- Manual administrator-driven assignment

You may have a more complex task assignment and reassignment system based on your specific business logic.

The main comment that I will add here is that this should be based on your specific business needs, and there is no "right" or "wrong" way to do it. The default is usually "on demand."

Do You Need to Customize the Interface?

Most tools assume you will customize the schema. Some also allow you to customize the look and feel of the UI, such as how large elements appear or where they are positioned. Others adopt a "standard" UI, similar to how office suites have the same UI even if the contents of the documents are all different.

You might want to customize the interface in order to embed it in the application, or you might have special annotation considerations.

Most tools assume a large-screen device, like a desktop or laptop, will be used.

How Long Will the Average Annotator Be Using It?

When considering tools, we have to consider not only how well they do what we want them to do, but when it comes to human tasks, we also have to think about what it would be like to do something repeatedly for long stretches of time. A simple example is hotkeys (keyboard shortcuts). If a subject matter expert is using the tool a few hours a month, then hotkeys may not be all that relevant. However, if someone is using it as their daily job, perhaps eight hours a day, five days a week, then hotkeys may be very important.

To be clear, most tools offer hotkeys, so that specific example is likely not worth worrying about, but it's a great example nonetheless. More generally, the point is that, by accidents of history or intent, most tools are really optimized for certain classes of users. Few tools work equally well for both casual and professional users. Neither is right or wrong; it's just a trade-off to be aware of.

Tasks and Project Structure

First, let's wrap our heads around the general organization structure. A project contains multiple tasks, as shown in Figure 5-2.

As you work with tasks you will use analytics functions, data exploration, model integration, quality assurance, and data flow concepts to control work in progress. A well-structured project makes work in progress more manageable. Usually, work is organized at the template level, and the individual tasks are generated by the training data system.

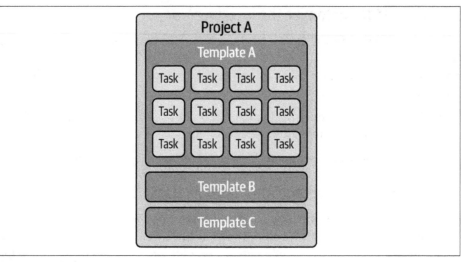

Figure 5-2. Task structure

Quality Assurance

Often, for people new to training data, there is an assumption that quality assurance surely can't be that hard.

I mean, just look at the data, right?

The main issue is that the amount of data a single administrator can see is typically a small fraction of the overall annotated set. In some cases, even a small sampling of the annotated data may be more than an admin could possibly review in their lifetime. This is a strangely hard fact for many people to accept. It seems so easy to look at "a few images."

As we can see in Figure 5-3, the amount of data a person can see in a reasonable amount of time is often a tiny fraction of the overall dataset size. Often, in practice, the difference is many orders of magnitude greater. There's simply too much data!

The solutions to this problem generally involve getting more people involved and developing more automations. We will cover automations in a later chapter, so for now, we'll concentrate on the former solution. With more people involved, the next big question is really "How much do I trust my annotators?"

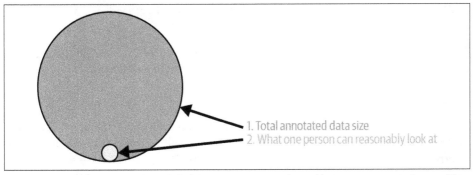

1. Total annotated data size
2. What one person can reasonably look at

Figure 5-3. Comparison of total dataset size compared to what a person can reasonably look at in a reasonable time frame

Annotator Trust

Many early efforts in computer vision generally assumed that annotators were "heathens." Essentially, data scientists thought that they couldn't be trusted, at least as individuals. This led to a lot of research on verifying the wisdom of crowds. At a high level, this boils down to the idea that accurate results come from having multiple people do the same task and then analyzing the aggregated result. But is this the best approach? Let's take a closer look.

Annotators Are Partners

Annotators are a key part of the QA loop. Often, I have seen people jump to an "us versus them" mindset when it comes to quality assurance. While annotators need to understand QA goals and their importance, and that rigor is especially impactful in larger projects, we are all on the same team here.

I think that often, concerns regarding annotator agreement and quality are overstated. For many of the annotation issues that arise, people do generally agree with each other. Often, when there is disagreement, the schema, resolution of data, hidden assumptions, etc., are more to blame than any single annotator. Working in partnership to define the specs of the project before it begins and building feedback loops into your process will provide the highest quality and most accurate results.

Who supervises the data

There are psychological elements, human backgrounds, the context of annotation, personal goals, etc., that are well beyond the scope of this book, but it's worth considering broadly that there are *humans* behind the data here.

As an analogy, consider that a programmer is involved in the compilation of her code. At the end of the day, it runs or it doesn't. She gets feedback on her efforts. Whereas here, an individual person may never get the same degree (or speed) of feedback on their annotation efforts. And they may never get feedback on their effect on the ML models.

Think of it almost like doing a "code review"— with all the effort we go through in a formal code review to protect feelings, emotions, and nascent ideas.

All training data has errors

Training data quality is of critical importance, the same way code quality is. Similarly, all computer programs have bugs and the same way, all training data has some form of errors. Annotators are only one source of errors. From what I have seen, it's best to focus energy on systemic issues and processes over "one-offs." ML models are surprisingly robust to occasional bad training data. If there are 100 "truly correct" examples, and 5 "misleads," in general the model will still work as expected.

For the cases where it really, really matters that 100/100 are correct, I think either we are kidding ourselves with that goal, or if it does happen, it's rarely. Yes—if a machine learning model is landing the Mars mission, OK, sure, the training data needs to be triple-checked. But otherwise, we are being penny wise and pound foolish trying to over-optimize every "error" out of the system.

Annotator needs

People's psychology and general work ethic is often at least as important as any specific QA process. People who are respectfully treated, compensated, and trained, will naturally perform differently than those subject to "lowest bidder" approaches. Try to put yourself in their shoes, and consider what they might need to succeed, and how they are actually encountering the data. There are a few key questions for managers to keep in mind when addressing staffing:

- Who is actually annotating the data?
- Are they trained on the tooling? Common expectations (such as zooming in) may not be clear, even to professionals, unless explained. Don't assume!
- What other kinds of training have they received?
- Is their technical setup appropriate? Are they seeing the same thing I'm seeing? (For example, common screen resolutions can differ between countries.)
- Do they have an individual account where we can track their specific concerns and performance metrics?
- Are the performance metrics an accurate reflection on the person?

Video Meetings Example

As a quick aside to consider how knowledge of training data and tooling does not correlate with annotator intelligence, consider video meetings. I know when I join a video meeting with a platform I'm unfamiliar with, even simple operations like muting and unmuting can be hard to find, or entirely neglected. Annotation is similar. One tool may do something easily, or even automatically, whereas another might not. Don't assume that subject matter expertise equals expertise using a tool. Even a doctor who has 30 years' experience with cardiology, say, may not know that a specific tool has a specific feature.

Common Causes of Training Data Errors

There are several common causes of training data errors to keep in mind:

- Raw data is misaligned for the given task (e.g., for computer vision if the data is low resolution and pushing the limit of what a human can reasonably see).
- The overall schema is suboptimal; e.g., a human annotator cannot reasonably represent their knowledge in the schema or a model cannot reasonably predict the schema.
- The guide is suboptimal or contains errors, such as contradictions.
- Specific schema attributes are unreasonable (e.g., too specific, overly broad, or misaligned with raw data).
- Labels and attributes are free-form (annotators enter own novel text/label).
- Annotators are working with more complex spatial types, or overly specific spatial types (pixel specific).
- Any form of "hidden" information, e.g., occlusion.
- Any naturally controversial content (e.g., politics, free speech/hate speech, facial recognition).

Task Review Loops

A task review is when another person looks at the previously "completed" task. Two of the most popular review options are the standard review loop and consensus.

Standard review loop

Every task has some percent chance of being sent for review. If it is sent for review and rejected, it will go back to "in progress," as shown in Figure 5-4. If it is then accepted, it will be considered "completed." A UI example of this is shown in Figure 5-5. To sum up, the loop has four components:

---

1. In progress.

2. In review.

3. If fail review, go back to in progress.

4. If pass review, complete.

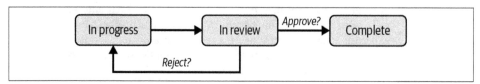

Figure 5-4. Diagram of review loop flow

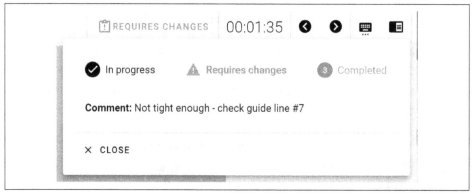

Figure 5-5. Diagram of changes required

Consensus

In the context of annotation, consensus is the word used to describe aggregations of multiple judgments about a particular sample with a common one produced through analysis. For example, in Figure 5-6, we see three people drawing a similar box, and the combined result is analyzed into one result. If it sounds a little overwhelming to have multiple people do the same thing and try to combine the results, you're right! It is! Generally, it at least triples the costs involved. It also introduces a bunch of challenges around analyzing the data. Personally, I also think it leads to the wrong type of analysis. It's generally better to simply get more samples, supervised by different people, than to try to get some grand conclusion on single samples. For example, if there's wild disagreement—what do you do?

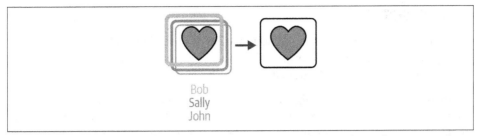

Figure 5-6. Example of consensus approach showing overlapping boxes being merged into one

As a rough analogy, imagine asking three engineers to each write an application independently and then algorithmically merging the three versions together—it just doesn't make sense! Compare that to the three engineers working on different sections, then coming together to discuss concerns on specific design trade-offs.

The second major problem this introduces is that most real-world cases are much more complex than a single instance. Attributes, video, etc.... If one person says it's frame 12, and another one says it's frame 13—do you split the difference? While there's some very real research in this area, generally, as the complexity of the challenge has increased, and the "seriousness" of the annotators has increased, the need for consensus has decreased.

Analytics

Like with any task management system, reporting and analytics are key to understanding the performance of annotators and subject matter experts. It can also be used to analyze the datasets themselves, as well as schema and other system components.

Annotation Metrics Examples

There are some common metrics that will help you assess your annotation:

- Time per task. Cautions: tasks can vary substantially. This metric can be very dangerous to use by itself; suggest combining with quality metrics.
- Number of "accepted" tasks. Accepted could mean passed review, was not reviewed, or similar metric.
- Count of instances updated or created.
- Count of instances per task or per frame.
- Count of tasks complete. Also consider the difficulty of tasks and type of tasks.

Data Exploration

To some extent, exploration can be thought of as a "supercharged" version of looking through files on a regular file browser. Typically, this means with the aid of various tools designed for this specific domain.

As an analogy, consider looking through marketing contacts in a marketing system versus a spreadsheet. You may still see the same "base" data in the spreadsheet, but in the marketing tool it will provide other linked information and offer actions to take, such as contacting the person.

Exploring data includes manually writing queries and viewing it. It can also include running automatic processes that will walk you through the data, such as to discover insights or filter the volume of it. Exploration can be used to compare model performance, catch human errors (debug the human), and more.

While it's possible to do some limited exploration by manually looking at raw data in a regular old file browser, that's not really what I'm referring to here.

To explore the data, it must have already been ingested into a training data tool. Data exploration tools have a lot of variance, especially around proprietary automatic processes and depth of analysis.

The data can often be viewed as a set of samples or a single sample at a time.

Here are important considerations when reflecting on exploration:

- The person who does the exploration may, or may not, be involved in the annotation process. More generally, the person doing the exploration may be different from the person who did any other process, including uploading, annotation, etc.
- The organization and workflow of data to do annotation may not be useful to someone using the data for actual model training.
- Even if you are directly involved in all of the processes, annotation of the data is separated by time from exploration of the data. For example, you may conduct workflow for annotation over a period of months, and then on the third month do further exploration, or you may explore the data a year later, etc.

To put this concretely, if I'm concerned about annotation, I care about "What's the status of 'batch #7'?"

When I'm exploring the data, I may want to see the work of all batches from 1 to 100. At that point, I don't necessarily care what batch created it; I just want to see all examples of some label. Put more broadly, it's in part a different view of the data that crosscuts multiple datasets. Put simply, the exploration process may be separated by time and space from annotation.

Exploration can be done at virtually any time:

- You can inspect data prior to annotation of a batch, such as to organize where to start.
- You can inspect data during annotation to do quality assurance, to simply inspect examples, etc.

Generally, the goals are to:

- Discover issues with the data.
- Confirm or disprove assumptions.
- Create new slices of the data based on knowledge gained in the process.

Data exploration tool example

A training data catalog can allow you to:

- Access a slice of the data instead of having to download all of it.
- Compare the model runs.

Explore processes

- Run a query, filter, or program to slice or flag the data.
- Observe the data.
- Select or group the data.
- Take some action.

Explore examples

- Flag a file or set of files for further human review, e.g., for missing annotations.
- Generate or approve a new novel slice of the data, e.g., a reduced dataset that may be easier to label.

Let's take a deeper dive into the example of similar image reduction.

Similar image reduction

If you have many similar images, you may want to run a process to reduce them to the most interesting ten percent. The key difference here is that often it's an *unknown* dataset, meaning there are few or no labels available yet. You may know that 90% of the data is similar, but you will need to use a process to identify which images are unique or interesting to label.

Once you have your data in your training data tooling as the first ingest step, it becomes easy to add an image reduction step from there. Really, this step is automatically exploring the data, taking a slice of it, and then presenting that slice for further processing.

Often the organizational methods necessary during the creation and maintenance of training data are less relevant to the creation of models. The people who create the training data may not be the people who create the datasets. And again, others may take the sets and actually train models.

Models

Working with ML models is part of your training data workflow.

Using the Model to Debug the Humans

A key additional aspect to importing data is tagging which "model run" the instances belong to. This is to allow comparison, e.g., visually, between model run(s) and the ground truth (training data). It can also be used in quality assurance. We can actually use the model to debug the training data. One approach to this is to sort by the biggest difference between the ground truth and predictions. In the case of a high-performing model, this can help identify invalid ground truths and other labeling errors. These comparisons can be between the model and the ground truth, between different models exclusively (no ground truth), or other related combinations.

Figure 5-7 shows an example of a model prediction (solid line) detecting a car where the ground truth was missing it (dashed line). An example algorithm is comparing the nearest Intersection over Union (IoU) to some threshold; in this case, it would be very high since the box hardly overlaps with any of the light-colored ones. This type of error can be automatically pushed to the top of a human review list because the box is far from any others.

Figure 5-7. Model predictions being compared to human predictions

To further help picture the relationships here, consider that the ground truth changes at a slower frequency than the model predictions. Sure, errors in ground truth may be corrected, more ground truth added, etc. But for a given sample, in general, the ground truth is static. On the other hand we expect during development that there will be many model runs. Even a single automatic process (AutoML) may sample many parameters and generate many runs, as shown in Figure 5-8.

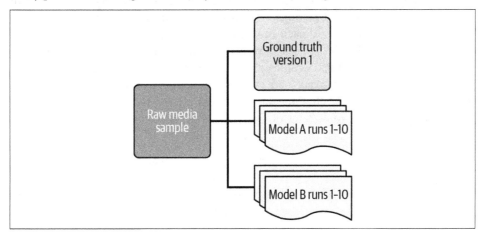

Figure 5-8. One file may be relevant to many model runs and ground truth sets

There are a few practical notes to keep in mind here. There is no requirement to load all the model predictions back into the training data system. In general, the training data is already used in a higher-level evaluation process, such as determining the accuracy or precision, etc. For example, if there is an AutoML process that generates 40 models and identifies the "best" one, you could filter by that and only send the best predictions to the comparison system.

Similarly, an existing human-supervised dataset is not strictly required, either. For example, if a production system is predicting on new data, there will not be ground truth available. It can still be useful with a single model to visually debug it this way, and it gives flexibility for other cases. For example, you might want to ship a new version of the model and be able to spot-check it by running both old and new versions in parallel during development efforts.

Distinctions Between a Dataset, Model, and Model Run

There is no strict relation or hierarchy to a "dataset" and a "model"—it's better to think of their relationship on a per-sample level. For example, Sample 1 and Sample 2 may be in the same set, but model version 1 may have been run on Sample 1 only. A model run may generate a *set* of predictions, but the *real* dataset is the amalgamation of *all* of the predictions, as well as the ground truth. This overall structure is defined more by other needs than by what sample or batch of samples happened to have been run through a model. While in some contexts this difference may be mincing words, in my opinion, it's best to reserve the "training dataset" concept in this case for a real, "complete" dataset, not a partial artifact the model produces as part of its process.

The "model" itself is typically understood to have the form of the raw weights, which are also known as the direct output of a training process, whereas a model run adds context, such as what settings (e.g., resolution, stride) were used at runtime. Therefore, it's best to think of each model run as a unique ID. This is because it is, of course, a model with the same structure, but trained differently (data, parameters, etc.). And a static model run is still unique, because the context (e.g., resolution or other prep) may have changed, and the model "runtime" parameters (e.g., stride, internal resolution, setting, etc.) may have changed. This can get context-specific very quickly. In general, it avoids a lot of confusion to have a unique ID for each set of {model, context it runs in, settings}.

Model runs are also known as predictions. A machine learning model can be run on a sample or dataset. For example, given a model X, with an image Y input, a prediction set Z should be returned. In a visual case, this could be an object detector, an image of a roadway, and a set of bounding box type instances.

Getting Data to Models

One of the big ideas is that of the perpetual improvement of models. This is accomplished by repeatedly updating training data and updating models. In the ideal case, this is done through a "streaming" type of mechanism, e.g., a new model prediction is automatically pushed to a review system if criteria are met. We will talk more about the "MLOps" process later. For now, the main thing to think about is: Where do I want the data to go after that task is completed? What kind of stages do I want to do? Next, I will cover dataflow, which includes getting data to the models.

Dataflow

There are two major concepts to be aware of when it comes to how data flows through your system. One is the overall workflow, including ingestion, tasks, model training, etc. The other is task-specific dataflow. In this section, I will zoom in to talk about dataflow only.

You are likely to have many pipelines for different contexts. For example, you may have a training data pipeline—that's for people-oriented tasks. Then there may be a model pipeline—that's how the training process works, how the best models get to production, etc. And throughout all of that, there may be some kind of generic pipeline, e.g., Apache Airflow, that moves data from some other system to the training data system in the first place. We have discussed concepts around perpetually improving models, but how do we actually achieve it? There are a few ideas at play here; I'll unpack each.

Overview of Streaming

The 10,000-foot-view goal of streaming is to automatically get human annotations "on demand." While people often jump to thinking of real-time streaming, that's actually a different (albeit related) idea. Instead, think of a project where one person on the team has defined the label schema. But the data is not yet ready—perhaps because the engineer who needs to load it hasn't loaded it yet, or perhaps because it's novel data that has not yet been captured by a sensor. While those sound very different, from the training data perspective, it's the same problem: the data is not yet available.

The solution to this is to set everything up in advance, and then have the system automatically generate the "concrete" tasks (from the template configuration) when the data becomes available.

Data Organization

Like any data project, there is organization required for the dataset itself, separate from the human task completion. Common approaches include:

- Folders and static organization
- Filters (slices) and dynamic organization
- Pipelines and processes

Folders and static organization

For computer data, I picture files and folders on a desktop. Files that are organized into folders. For example, if I put 10 images in a folder "cats," I have in a sense created a dataset of cat images. This set is static, and doesn't change based on a query or event.

Filters and dynamic organization

A dataset may also be defined by a set of rules. For example, I could define it as "All images that are less than six months old." And then I could leave it to the computer to dynamically create the set on some frequency of my choosing. This overlaps with folders. For example, I may have a folder called "*annotated_images*," the contents of which I further filter to show only the most recent x months.

Pipelines and Processes

These definitions may also become more complex. For example, medical experts have a higher cost than entry-level people. And running an existing AI has a lower cost still. So, I may wish to create a data pipeline that goes in the order: AI, entry level, expert.

Arranging tasks purely by date would not be as useful here, since as soon as the AI completes its work, I want the entry-level person to be able to look at it. It's the same when the entry-level person completes their work; there's always a next step in the process that should get started as soon as possible.

At each stage of the process, I may wish to output a "folder" of the data. For example, say that we start with 100 images that the AI sees. Say that at a given moment in time, an entry-level person has supervised 30 images. I may wish to take just those 30 images and treat that as a "set."

In other words, the stage in the process it's at, the status of the data, and its relationship to other elements helps determine what constitutes a set. To some extent, this is like a hybrid of folders and filters, with the addition of "extra information" such as completion statuses.

The implementation of pipelines can sometimes be complex. For example, the label sets may be different.

The dataset connection

How do we know when the data is available? Well, first we need to send some kind of signal to alert the system that new data is present, then route the data to the right template.

Let's jump to code for a moment to think about this. Imagine I can create a new dataset object:

```
my_dataset = Dataset("Example")
```

This is an empty set. There are no raw data elements.

Sending a single file to that set

Here I create a new dataset and a new file, and add that file to the set:

```
dataset = Dataset("Example")
file = project.file.from_local("C:/verify example.PNG")
dataset.add(file)
```

Relating a dataset to a template

Next, I create a new template. Note that this has no label schema; it's an empty shell for now. Then I have that template "watch" the dataset I created. What this means is that every time I update a file to that set, that action will create a "callback" that will trigger task creation to that set automatically:

```
template = Template("First pass")
template.watch_directory(my_dataset, mode='stream')
```

Putting the whole example together

Here, I created a new template (for humans), and the new dataset where I plan to put the data. I instruct the dataset to watch for changes. I then add a new file to the system—in this case an image.

Note that at this point, the file exists in the system, in whatever default dataset is there—but in the dataset that I want. So in the next line, I call that dataset object specifically and add the file to it, thus triggering the creation of a concrete task for human review:

```
# Construct the Template
template = Template("First pass")
dataset = Dataset("Example")
template.watch_directory(dataset, mode='stream')

file = project.file.from_local("C:/verify example.PNG")
dataset.add(file)
```

 Practically speaking, many of these objects may be a .get() (e.g., an existing set). You can target a dataset at the time of import (it doesn't have to be added separately later). These technical examples follow the Diffgram SDK V 0.13.0, which is licensed under MIT open source approved license.

Expanding the example

Here I create a two-pass template. The "two" because the data will first be seen by the first template, and then later by the second template. This is mostly reusing elements from the prior examples, with the upon_complete being the only new function.

Essentially, that function is saying "whenever an individual task is complete, make a copy of that file, and push that to the target dataset." I then register the watcher on that template like normal:

```
template_first = Template("First pass")
template_second = Template("Expert Review")
dataset_first = Dataset("First pass")
dataset_ready_expert_review = Dataset("Ready for Expert Review")

template_first.watch_directory(dataset_first , mode='stream')
template_first.upon_complete(dataset_ready_expert_review, mode='copy')

template_second.watch_directory(dataset_ready_expert_review, mode='stream')
```

There is no direct limit to how many of these can be strung together—you could have a 20-step process if needed here.

Non-linear example

Here I create three datasets that are watched by one template. The point here, as far as organization goes, is to show that while the schema may be similar, the datasets can be organized however you like:

```
template_first = Template("First pass")
dataset_a = Dataset("Sensor A")
dataset_b = Dataset("Sensor B")
dataset_c = Dataset("Sensor C")
template_first.watch_directories(
[dataset_a, dataset_b, dataset_c], mode='stream')
```

Hooks

For complete control of this process, you can write your own code to control this process at different points. This can be done through registering webhooks, userscripts, etc.

For example, a webhook can be notified when a completion event happens, and then you can manually process the event (e.g., by filtering on a value, such as a number of instances). You can then programmatically add the file to a set. (This essentially expands on the copy/move operation upon_complete().)

An example of how to achieve the upon_complete() in a user interface is shown in Figure 5-9.

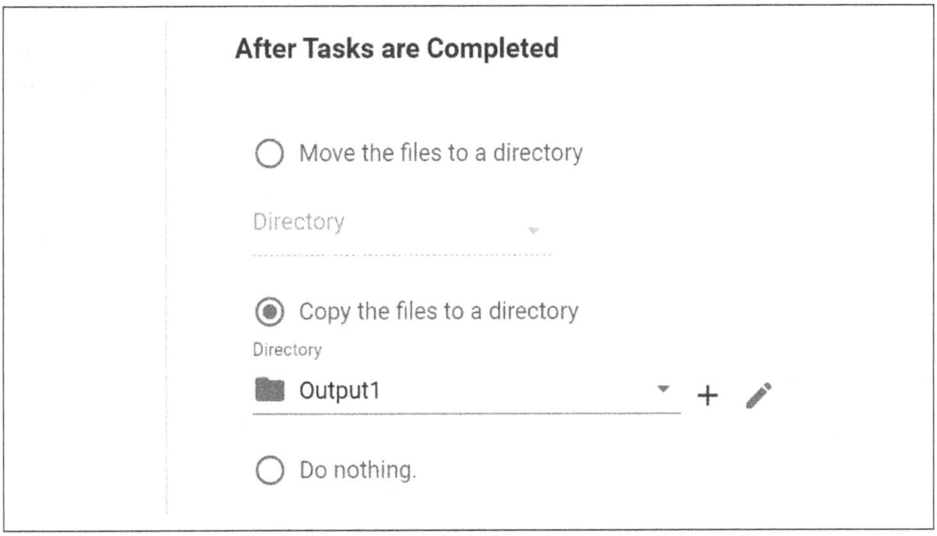

After Tasks are Completed

○ Move the files to a directory

Directory

● Copy the files to a directory
Directory
📁 Output1   ▾ + ✎

○ Do nothing.

Figure 5-9. Task completion UI example

Direct Annotation

This section is about the practice of annotation, but it's recommended reading for admins too. It's absolutely valuable for administrators to know what an annotator's experience is. Meanwhile, an annotator looking to lead a team, move into an administrator role, or who is simply curious about the bigger picture can benefit from this section.

As an annotator, you are supervising data (writing the code) that powers AI. You have a huge responsibility to map noisy real-world data onto a schema defined by others. On our journey, we will cover some core concepts of actually doing annotation, including video series, images, and mechanics.

Annotators have a ground-level view of the data. This means you will often have valuable insights to contribute about the effectiveness of the schema relative to the actual data. If you are a subject matter expert, you may have additional efforts to lend in setting up and maintaining the schema. Often, by the time it gets to actually annotating, the initial schema has been defined. However, there's still lots of opportunity to help maintain and update it as issues are encountered, such as the bike rack versus bicycle example we covered earlier.

Every application has specifics, such as buttons, hotkeys, order of steps, and more, that are slightly different. UIs will naturally evolve and change over time too. Some media-type UIs may be intuitive or easier to learn. Others may require substantial training and practice, to master.

An important note is that this section focuses in on a small slice of complex annotation. There is often great value added in even simple annotations done by the right people at the right time, so your project might not require careful study of all the details in this section.

Next, I will cover the two common annotation complexities, supervising existing data and interactive automations. Then I will ground all of this by taking a deeper dive into video annotation. While there are many media types, and UIs are always changing, video gives a good example of something that is more complex to annotate well.

Business Process Integration

An emerging option is to frame existing day-to-day work as annotation. By reframing a line of business-specific workflow as annotation, you can get annotations for little extra cost. In general, this requires being able to configure the annotation UI in a way that can match the business workflow. An annotation may both complete an existing business process and create training data at the same time.

Attributes

Annotation of attributes continues to increase in complexity. Essentially, anything you can imagine being placed in a form can be theoretically defined as an attribute. While historically, spatial annotation (where something is) has been a large focus, increasingly, attributes are becoming important. I have seen some projects with tens and even hundreds of groups of attributes.

Depth of Labeling

As shown in Figure 5-10, rendering a video and asking questions about the entire video is very different from frame-specific labeling. Yet, both often get shepherded under the term "video labeling." Explore the depth of support needed carefully for your use case. In some cases, a little more depth may be easy to add after, and in other cases it can completely change the overall structure of the schema.

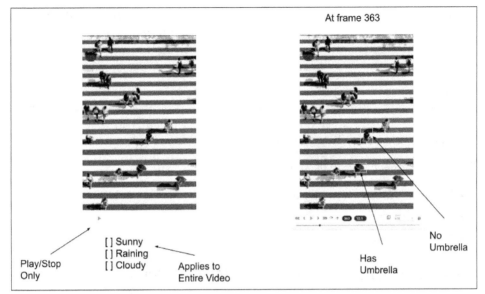

Figure 5-10. Depth of labeling—comparison of whole video versus frame (image from Unsplash (https://oreil.ly/Rc8id).

Supervising Existing Data

A popular automation approach is pre-labeling. This is where the model has already made a prediction. Depending on the use case, you may be asked to correct static automations, add more detail to them, or otherwise interact with them. An example process is to review the file, update it, and mark it complete.

Anytime you see a pattern, this is a great opportunity to help improve the model. Are you always correcting a similar seeming error? Communicating that to the admins or data science team can go a long way to improving the model. The pattern you found might be relevant even if the schema is correct, it could be something internally wrong with the model that needs fixing.

Interactive Automations

Supervising existing predictions is different from interactive automations, where you may be able to continually work with a more complex process, usually in real time, until some result is achieved.

An example of interactive automation is drawing a box around an area of interest and automatically getting a polygon drawn closely around the object. Another is clicking key points and getting segmentation masks.

More generally, an interactive automation is where you add more information into the system, and then some process runs based on the new information you added.

Sometimes this can be done iteratively until some desired state is reached. This can sometimes be a "try again"–type thing, where you keep trying to draw a box until it gets the right polygon, or it can be a "memory"-based system, where it continues to change based on your ongoing inputs.

Usually, as an annotator, you won't need to actually "code" any of these interactions. They will often be provided as UI tools or hotkeys. You may need to know a little about the operating parameters of the automations—when they work well, when they work poorly, when to use what automation if there are many options available, etc.

Some common methods to be aware of:

- "Prompt" (draw box or points, etc.) conversion to Polygon/Segmentation
- Full-file general prediction
- Domain-specific ones, like object tracking or frame interpolation

Example: Semantic Segmentation Auto Bordering

Let's zoom into a specific example, auto bordering, shown in Figure 5-11. This is where the system detects edges to create 100% coverage masks. This is generally faster and more accurate than trying to draw the border manually.

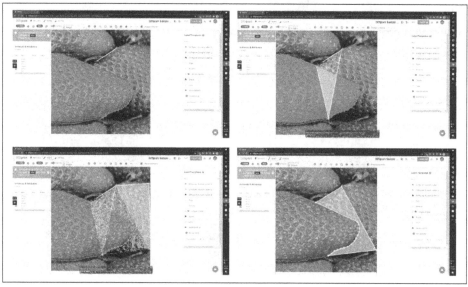

Figure 5-11. Example of UI showing auto bordering process

The steps are as follows:

1. Select a point on the intersecting shape.
2. Select the exit point on the intersecting shape.

OR

1. "Draw over," e.g., draw over an existing object and expect that it will auto border around the intersection of the points.

Video

The more complex a media form and schema are, the more a great UI/UX experience and automations can save time. What's maybe not a big deal in a single image review may be a massive problem in an 18,000-frame video. Now, let's cover some more complex examples with video.

Motion

We annotate video to capture meaning in motion—cars moving, a basketball shot being taken, factory equipment operating. Because of this motion, the default assumption when annotating is that every frame is different, as shown in Figure 5-12.

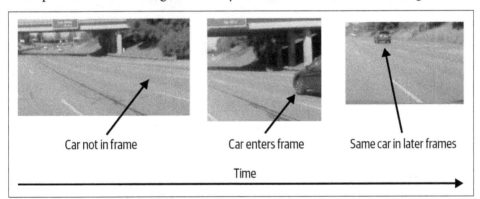

Figure 5-12. Frames showing over time the car is not in view, then enters, and the same car appears in different frames

As you can see in Figure 5-12, the car is not in the initial frame. It moves into the frame. Then later in time, it is still in the frame. More generally, objects move in and out of the frames. This happens at different points in time.

Attributes of the car may change too—for example, in one frame it may be fully visible, and in the next it's partially occluded.

Examples of tracking objects through time (time series)

The goal of a time series is to create some relation between annotations from multiple points in time (often frames or timestamps). Time series are also known as sequences, tracks, and time consistency.

Examples of ways to track objects through time in the UI include the following:

"Ghost frames"
Where the previous position is marked and users move that frame to represent the current state

"Hold and drag"
Where a user holds onto an object and moves it as the object moves

"Point and predict"
Where a user marks key points and a tracking algorithm guesses at the movement

One approach is to create (or select) a series. Each series can be unique within each video, unique to a set of videos (e.g., the same scene taken with multiple cameras), or globally unique across videos. Generally, this forces every object to be part of a sequence. This works well if objects are generally present in more than one frame, but it can be a little bit cumbersome if objects are routinely only in one frame.

Generally, this series/sequence approach enforces additional constraints:

- A sequence can only appear once in a given frame. For example, series 12 cannot appear twice in the same frame. This may not always be correct—for example, an object may be partially occluded and could be represented by two or more spatial types. (Picture a bus blocked by a post.)

- A sequence must be of the same label type. While an attribute may change, the "top level" concept should generally be the same between frames.

Static objects

Sometimes, a video will have a static object, e.g., a retail store with a shelving unit or other display that doesn't move (or doesn't move often), or an intersection with a traffic light that's not moving. You can represent this in three ways:

- A single keyframe, i.e., at frame x (e.g., #898).

- A keyframe at the start and end (e.g., 9, 656). This means that the object enters at frame 9 and exits at frame 656. More generally, this pattern is (entry frame, list of frames, exit frame).

- An attribute, such as "enters" or "visible" (or "% visible").

These are all ways to effectively tag something as a "static object."

Persistent objects: football example

A video may have multiple objects in it. For example, there can be two different cars, several apples, football players running on a field, etc.

From the human perspective, we know a football player, such as the player in Figure 5-13, is the same person, in frame 0, 5, 10, etc. In every frame, he is still Ronaldo.

Figure 5-13. A football player in a frame (Source: Unsplash (https://oreil.ly/7P8HV))

But this is not as clear to a computer. So to help it, we create a sequence object, i.e., "Ronaldo." Since it's the first object we created, it gets assigned sequence #1. If another player, "Messi," was also in the frame, we could create a new sequence for him, and he would get #2. Another player would get #3, and so on.

The key point is that each sequence represents a real-world object (or a series of events), and it has a number that's unique for each video.

Series example

We can also create a set of "series" to create meaning over time. Imagine a video with three objects of interest. To represent this, we create three series. Each frame represents an instance:

- Series #1 has frames (0, 6, 10) because Object #1 enters the video at frame 0, something about it changes at frame 6, and then it leaves the video at frame 10. The object is changing over time in each keyframe.
- Series #2 has frames (16, 21, 22). Object #2 also changes position from frame to frame.
- Series #3 has frame (0) only. Object #3 is a static object that doesn't move.

A series may have hundreds of instances.

Video events

There are a few ways to represent events:

- You can create a new series for each event.
- You can create a new frame for each event. For example, (12, 17, 19) would mean three events, at frames 12, 17, and 19.
- You can use attributes to declare when an "event" occurs.

From a UI standpoint, the main trade-offs here have to do with how many events you expect and how complex the rest of the schema is. As a rule of thumb, if there are fewer than 50 events, it *may* be "cleaner" to keep them all as separate series. If there are more than 50 events, it's generally better to have a single sequence, and use either frames or events.

Attributes can work well for complex event cases. The main downside is that they are often less "discoverable" than keyframes, which can be more visible to help you "jump" to a specific point in the video.

In terms of annotation speed, pay close attention to hotkeys for video events. Often, there can be hotkeys to create new sequences, change sequences, etc.

In terms of common annotation errors, be sure you are on the desired sequence. Often, thumbnails will help visually determine this. You can also jump back or forward to other frames in the sequence to verify where you are.

Detecting sequence errors

Imagine you are reviewing a video that was pre-labeled (either by another human or an algorithm). To aid with this, tools like Diffgram automatically change the color of the sequences. This means you can play the video, and watch for color changes in the sequence. It can be surprisingly easy to catch errors this way. An example of an error is shown in Figure 5-14. See how the vehicle's sequence number changes even though it's the same vehicle. Because it's the same vehicle, it should have the same sequence number over time.

For real-world cases with overlapping instances, often playing the video with this color feature is the easiest way to detect issues. You also may be alerted to a potential series error through an automated process or another annotator raising a ticket on it.

Figure 5-14. Example where the same vehicle's sequence number changes, incorrectly

To correct the error, I open the context menu of the instance and select the correct sequence, as shown in Figure 5-15.

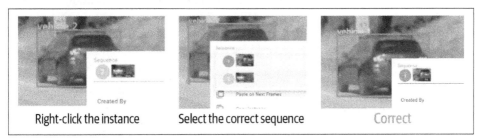

Right-click the instance   Select the correct sequence   Correct

Figure 5-15. An example of the process to correct an invalid sequence

In this example, I know it's correct because of two things:

- The car is most similar to its prior frame.
- The car is visually similar to the thumbnail of the series.

Common issues in video annotation

We, as humans, may take in a given scene, such as a driving roadway, and assess where the shoulder of the road/vegetation is. However, when purely looking at the image pixels, there is not significant visible evidence for it! More generally, this is about declaring what "should" exist versus what you can actually see. Other common issues include:

- Reflections, e.g., a human's reflection in a window.
- Objects that create barriers but are see-through, e.g., railings.
- Object detection predicts a rectangle shape, whereas most real-world objects are not rectangles.
- Busy areas out of driveable view, e.g., the Las Vegas strip.
- Issues of objects appearing and disappearing in video frames.

Summary

Workflow is the glue between technology and people in training data. It includes the human tasks, as well as the surrounding automation and dataflow. Business workflows may be framed as or incorporate human labeling to get great quality annotations while completing normal work. Overall, these human tasks are the core building blocks of workflow. Tasks require schemas, permissions, and dataflow. Additional steps like review loops and analytics provide quality assurance.

Common causes of annotation errors include sensor or resolution limits, suboptimal schemas, inconsistent guides, and complex spatial types. Review loops and other tools reduce errors; however, building trust with annotators as partners often yields the best results. It's common to "blame the annotators," but often there is more leverage in changing the overall data structures, including changing the schemas.

Schemas require great up-front effort and expand over time, and changes often invalidate prior work. Therefore, careful consideration and understanding of the expected expansion directions is critical. This can also help keep schema leverage available to reduce annotation errors and improve performance.

There are many forms of complex annotation, including video. Video annotation captures motion over time, with techniques like ghost frames available for consistency through time. Interactive automations like auto-segmentation reduce repetitive work. Supervising existing predictions and automations boosts efficiency.

Analytics provide insight into annotator performance, data distribution, and model comparisons. Data exploration tools allow slicing datasets for issues discovery and creating new views. Different user roles have different permissions, with most users being annotators. Training helps set expectations, and Gold Standard data can be used for testing.

Pre-labeling (predictions) can reduce annotation workload and help debug training data issues through comparisons to other predictions and ground truth. Each model run should be uniquely identified with its runtime settings. Dataflow automates annotation by streaming new data to human tasks. Hooks allow custom coding at each dataflow step.

Overall, workflow stitches together the human tasks and technical infrastructure to enable useful training data creation. Careful design and iterations of the workflow improve productivity and model performance.

CHAPTER 6

Theories, Concepts, and Maintenance

Introduction

So far, I have covered the practical basics of training data: how to get up and running and how to start scaling your work. Now that you have a handle on the basics, let's talk about some more advanced concepts, speculative theories, and maintenance actions.

In this chapter I cover:

- Theories
- Concepts
- Sample creation
- Maintenance actions

Training a machine to understand and intelligently interpret the world may feel like a monumental task. But there's good news; the algorithms behind the scenes do a lot of the heavy lifting. Our primary concern with training data can be summed up as "alignment," or defining what's good, what should be ignored, and what's bad. Of course, real training data requires a lot more than a head nod or head shake. We must find a way to transform our rather ambiguous human terminologies into something the machine can understand.

A note for the technical reader: This chapter is also meant to help form conceptual understandings of the relationships of training data to data science. The data science technical specifics of some of the concepts brought up here are out of the scope of this book, and the mention of the topics is only in relation to training data, not an exhaustive account.

Theories

There are a few theories that I think will help you think about training data better.

I'll introduce the theories here as bullet points, and then each one will be explained in each section:

- A system is only as useful as its schema.
- Intentionally chosen data is best.
- Human supervision is different from classic datasets (e.g., anomaly detection).
- Training data is like code.
- Who supervises the data matters.
- Surface assumptions around usage of your training data.
- Creating, updating, and maintaining training data is the work.

A System Is Only as Useful as Its Schema

To illustrate this, let's imagine a "perfect system."

Let's define perfection as that for any given sample, say a street image, it will automatically detect the schema, such as "traffic light" or "stop sign" 100% of the time, without failure.

Is it perfect in any commercially meaningful sense?

Well, you probably saw this coming, unfortunately, our "perfect" system is not really perfect; in fact, it's likely not even usable in any commercial application.

Because after celebrating and the dust settles, we realize we not only want to detect the traffic light, but also if it's red, red left, green, green left, etc. Continuing this example, we go back and update our training data with the new classes (red, green, etc.). And again, we hit a problem. We realize that sometimes the light is occluded. Now we must train for occlusion. Oops—and we forgot night examples, snow-covered examples, and the list goes on.

No matter how well the initial system is planned, reality is complex and ever-changing. Our needs and styles change. The system will only ever be as useful as our ability to design, update, and maintain the schema.

An "ideal AI" is perfect at detecting the *abstractions we defined* (the schema). Therefore, the abstractions, the schema, matters as much as, or more than, the accuracy of the predictions. Note that for GenAI systems this schema is part of the alignment process.

It's worth pausing here and considering that even if the algorithm is *perfect,* there's still a need to understand the training data, so as to ensure the abstractions fed to the algorithm are what we desire.

So how do we achieve this training data understanding? Well, any driver can tell you that there are a variety of driving conditions to be aware of. In that same direction, we consider that subject matter experts from whatever field is relevant (doctors, agronomists, grocery clerks) are involved. Therefore, the more those experts can be integrally involved in the design process, including the label schema, the better you can align the abstractions to the real world. This leads into our next theory: who supervises the data matters.

Who Supervises the Data Matters

Who supervises the data matters. From the obvious need to have subject matter experts annotate domain-specific data, to institutional knowledge in companies, to stylistic preferences, the importance of human supervision comes up again and again when working with training data.

For example, for a grocery store, a store employee is likely able to identify what a strawberry looks like and if they are OK to sell or not (e.g., moldy), whereas a farm may require a farmer or agronomist. Both systems may detect strawberries, but with very different schemas and goals. Going back to the grocery store, perhaps store A may have a different threshold for what's OK to sell than store B.

This leads us to considerations around who supervises the data—their backgrounds, biases, incentives and more. What *incentives* are the people who supervise the data given? How much time is being spent per sample? When is the person doing it? Are they doing a hundred samples at once?

As a supervisor, you may work for a firm that specializes in data supervision, or you may be hired by one company, or you may be a subject matter expert. Finally, the supervision may come directly from an end user, who is potentially unaware they are even doing supervision, e.g., because they are doing some other action in an application that just happens to also generate training data. An example comparison between being aware of the supervision being done versus unaware is shown in Figure 6-1.

Generally, if an end user is doing the supervision, the volume and depth of supervision will be lower. However, it may be more timely, and more contextually specific. Consider that both can be used together. For example, an end user suggesting that something was "bad" may be used as a flag to initiate further direct supervision.

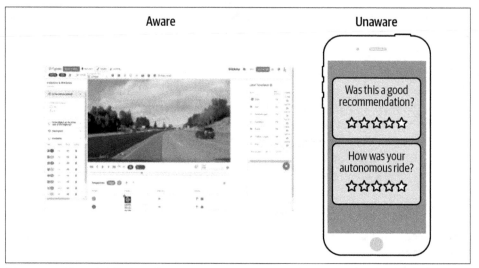

Figure 6-1. Visual difference between a scenario where the user is clearly aware they are training a system and one where the user may be unaware

This is all in the context of explicit (or direct) supervision of the data, someone is directly viewing the data and taking action. This contrasts with classic training data, where the data is implicitly observed "in the wild" and is not editable by humans.

Intentionally Chosen Data Is Best

Some companies scrape the internet to create big datasets from historical data. That's one approach; however, most supervised commercial datasets are novel. They consist of three main components:

- A newly created schema
- Newly collected raw data
- New annotations using the new schema and new raw data

Some examples of novel datasets are of privately recorded sports footage that is annotated, images of construction sites, and newly annotated sales calls. This is not random data from the internet. In fact, most supervised datasets are private (even if some of the original data may be public), not in the public domain.

Intentionally chosen data creates the best alignment between schema, raw data (BLOBs), and annotations. As we know, those must be well aligned to get good results.

Working with Historical Data

The opposite of fresh data is data from historical contexts. If the schema, raw data, or even annotations come from historical sources, they will naturally be bound or related to that historical context.

Why is this a concern? Historical data usually includes many problems, spanning from the technical to the political. For example, the historical data may have used a sensor different from modern sensors, or political concepts may have shifted such that the categories don't make sense anymore, etc. Historical data also may not have a well-defined "lineage," meaning that problems may not be obvious at first glance. Often, the very reason historical data is used is in an effort to save time or money. However, the larger the dataset, the more historical artifacts are encountered, and the more difficult they are to validate for modern contexts.

It's worth repeating here that in many classical machine learning problems, the historical data was the *primary* form of data available. For example, if you are training an email spam-detecting model, existing spam emails are a great source of data. However, in the supervised context, the domain space is often more complex, e.g., images, videos, and data derived from other sensors. The schema and prediction goals are then more complex. At the moment, the best way to achieve good results in this complex environment is to get great alignment between the schema, raw data, and annotations. Usually, overly leaning on historical sources makes this difficult.

You know you are creating "new" data and not bound to history by doing the following:

- Creating the schema yourself, or validating each record in an existing "must use" taxonomy, or reframing an existing taxonomy inside a new schema.
- Collecting novel data. It may be acceptable to also source data from existing datastores, as long as it's treated as raw data, and the age of the data is well understood.
- Annotating fresh data. This will always be required if either the schema or raw data are new, and ideally both are new.
- Any "must use" historical data is validated within the concept of the current situation.

It can be very tempting to lean into some historical concepts, especially for things like schema, where there may be well-defined existing taxonomies. If you must use an existing taxonomy, then validate as many records as you can for your new context. Ideally, for real-world cases, review as much of the data as possible, and it's especially important that the schema is created novelly, by you, for your specific needs. This means any framing concepts that could influence the analysis (bias, historical concerns, etc.) are reflections of your current view, not some generic historical concept.

To wrap up this section, I'll make a brief comment on costs. In Chapter 2, I already discussed the many cost trade-offs around creating new data. The bottom line when it comes to historical data is that it's often not really a true cost comparison of "use existing" versus "create new." Rather, the historical data is usually either technically or politically untenable at any realistic cost. When historical data is used, it's often only in part, and only when the parts that can be validated and brought forward are done so out of necessity or obvious usefulness.

Training Data Is Like Code

It's useful to think of training data like code because that helps frame the importance of it. The moment that someone says coding, I think "that's serious," and there should be the same reaction for the creation of training data. It's also useful in that it helps convey the importance of subject matter experts, and that they are exercising a very material level of control in the overall process. This helps frame the fact that data science is consuming the data produced by training data, not coming up with the "program" (model) exclusively by themselves.

Here, I will compare training data as a higher- level abstraction to already high-level programming languages like Python. Similarly to how we use languages like Python to be more expressive than we can be in assembly code,[1] we can use training data to be more expressive than we can using Python.

For example, instead of defining a program to detect strawberries using `if` statements, we can show an image containing a strawberry, then label a strawberry as the "desired result." This difference is illustrated in Figure 6-2. The ML process then uses the "desired result" to create a "program" (the model).

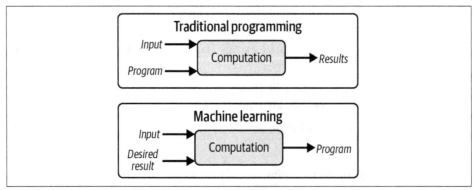

Figure 6-2. Difference between traditional programming and machine learning

---

[1] Assembly: "Any low-level programming language with a very strong correspondence between the instructions in the language and the architecture's machine code instructions." "Assembly Language" (*https://oreil.ly/SuBGz*), *Wikipedia*, accessed on September 14, 2023.

While conceptually this may be similar to how we just tell a child "that's a strawberry," in reality we are programming a specific technical label, "strawberry," and technical spatial coordinates, onto defined samples of data. While we may hope those samples are generalizable, the actual work done is the literal encoding of our knowledge into a dataset.

Software programs often have business logic, or logic that only makes sense in the business context exterior to the program. Similarly, training data relates to an external context, and is mapped to that exterior context of raw data and related assumptions. It's human meaning encoded in a form ready for consumption by a machine learning algorithm. Put simply, training data is another way to write programs via specifying desired results, rather than lines of code.

Shifting to a slightly more technical stance, training data controls a system by defining the ground truth goals. These goals eventually result in the creation of machine learning models. We may not know what model will be run on the data, but as long as the model outputs data similar to the ground truth, then the system is working as intended. This is similar to how we don't usually care how high-level languages like Python are "compiled/just-in-time compiled," as long as the code runs.

Of course, there are limits to this analogy, and there are many practical and philosophical differences. Code explicitly defines logic and operations, whereas training data provides examples that the model must generalize from. The intermediary step of model creation is less direct than that from code to program output. My intent is not to convince you there's a one-to-one analogy between coding and training data, but merely that training data is *like* coding in many forms.

Surface Assumptions Around Usage of Your Training Data

With training data, we define a set of assumptions, such as the schema itself, what the raw data will look like, and how the data will be used, and expect the training data to be useful in that context (and only in that context).

As an analogy, if you are an Olympic runner training to run the 100 meters, you are training to run based on the assumptions for the 100 meters, not the 400 meters, 800 meters, or the high jump. Because while similar, those events are outside of the scope of what you expect to encounter—the 100 meters.

Unlike the Olympics, where the conditions around the 100 meters are already well defined, with training data the assumptions are often not defined well, or even at all.

So, the first step is to start defining those assumptions. For example, is our strawberry picker assumed to be on a commercial strawberry field? Then, we get more specific. For example, perhaps we assume the camera will be mounted to a drone, or a ground-based robot, or the system will be used during the day, etc.

Assumptions occur in any system; however, they take on a different context with training data because of the inherent randomness involved.

The runner training for the 100 meters intuitively knows that the training for the 100 meters is different than for the 400 meters, 800 meters, or high jump. Getting a similar intuition around AI training is part of the challenge. This is especially hard because we typically think of computer systems as being deterministic, meaning that if I run the same operation twice, I will get the same result. The way AI models get trained, however, is not completely deterministic, and the world in which AIs operate is not deterministic. The processes around the creation of training data involve humans, and are not deterministic either.

Most training procedures have a variety of effects designed for randomness. For example, the initial model weights are usually random. Even if those weights are saved such that they can be reproduced, initially they were random. If all of the training parameters, such as data, model architecture, random seed value, training procedure, etc., are exactly recorded, it is generally possible to reproduce a similar model; however, that original model was still random.

Therefore, at the heart of training data is an *inherent randomness*. Much of the work with training data, especially around the abstractions, is defining what is and is not possible in the system. Essentially, we are trying to rein in and control the randomness into more reasonable bounds. We create training data to cover the expected cases, what we expect to happen. And as we will cover in more depth later, we primarily use rapid retraining of the data to handle the expected randomness of the world.

To end this section on a practical note, I'll list some common assumptions that need to be surfaced and reexamined when working with training data:

- Where, when, and how fresh data will be collected
- What the availability of retraining or updates to the schema will be
- What the "real" performance expectations will be, e.g., what is nice to have versus mission critical in the schema
- All of the annotation context, including guides, offline training sessions, annotator backgrounds, etc., that went into the literal annotation

Use definitions and processes to protect against assumptions

Defining a process is one of the most fundamental ways to set up guardrails around randomness. Even the most basic supervision programs require some form of process. This defines where the data is, who is responsible for what, the status of tasks, etc.

Quality assurance in this context is highly fragmented, with many competing approaches and opinions. We will discuss multiple opinions in the book. Generally, we will go from manual techniques toward more automated, "self-healing," multi-stage pipelines. An example process is depicted in Figure 6-3.

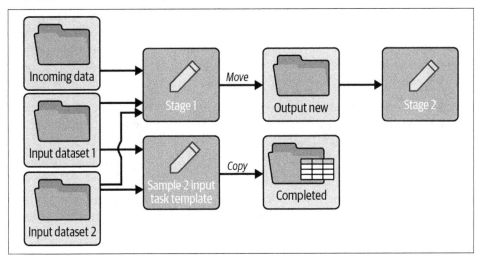

Figure 6-3. Visual graph of workflow in annotation process

As soon as you understand the basics of training data, you will quickly realize there is an obvious bottleneck: doing the literal annotation. There have been many approaches to speed up this core part of the process. We will explore the processes and trade-offs. Choices here can quickly become very complex. It's one of the most often misunderstood, but also one of the most important, parts of training data.

Human Supervision Is Different from Classic Datasets

"Supervised, semi-supervised, unsupervised"—these terms often come up in data science work. Any human intervention is some form of supervision, be it structuring the data, choosing features, or handcrafting a loss function. When I refer to "human supervised," I mean humans supervising the data, regardless of the technical approach used to model the data afterwards.[2] The general focus of this book is on the current understanding of supervised training data: the automation of what we already know, not discovery.[3]

---

2 For the technical reader, supervised learning is often thought of as the algorithm having access to the answer key (e.g., { input : output } pairs). While this is a valid definition, the focus here is on the human supervision aspect.

3 Training data can be used in combination with other approaches to discover new concepts in certain circumstances, but in general the bulk of the focus is on repeating existing known work.

Human-supervised training data is different from classic datasets. It's new. It has different goals, involves different skill sets, and uses different algorithmic approaches.

Discovery versus automation

In typical classical machine learning cases, we usually don't know what the answer is and we wish to *discover* it. In the new human-supervised case, we already know what's correct, and we wish to structure this understanding (produce training data) in a data science–consumable way. Of course, this is painting with a wide brush, and is not *always* true, but I use it to frame a contrast for an average case.

For example, in the classical case, we don't know what movie preferences someone has, so we wish to discover them, or we don't know what is causing a weather pattern and we want to discover it.

In a similar line of thinking, deep learning concepts are often treated as distinct concepts within machine learning. Often, human–supervised training data goes hand in hand with deep learning approaches because with deep learning there is an expectation of automatic feature extraction.[4] However, there is no strict requirement to use a deep learning approach.

As an example, a common relationship between the three concepts is shown in Table 6-1.

Table 6-1. Relation between goal, training data method, and algorithmic choice

Goal	Algorithmic approach example	Training data
Discovery	Machine learning	Classical
Automation	Deep learning	New—human-supervised

Discovery and automation goals are the two major complementary, and sometimes intersecting, camps within training data.

Consider the diagram in Figure 6-4. On the left is a spreadsheet, and on the right, a view of the world from a car. In the classic case, which is the spreadsheet, the data is already fixed. There is little to label. In contrast, in the new case, the raw data doesn't mean anything by itself. Humans must add labels to control the meaning. This means there are more degrees of freedom and more capabilities. In other words, in the classic context there is only indirect human control, whereas in the new context there is direct human control.

---

4 For the technical reader, as we know, deep learning is usually viewed as a subset of machine learning. I will sometimes refer to machine learning (or even AI) as a catch-all term to encompass things relevant to either approach. The intent of this is to maintain the focus on the training data art—how the data is modeled is up to data science.

Owner	Country	File_date	IPC_class
Company A	US	6/18/2008	H05H13
Company A	EP	1/30/1998	A61N5
Company A	EP	1/30/1998	A61N5
Company A	EP	1/30/1998	A61N5
Company A	JP	8/28/1997	A61N5
Company A	JP	10/4/2002	A61N5

Figure 6-4. Comparison of tabular data taken (discovery) as is versus complex human-supervised data

This is not to diminish the continued importance of classic datasets or discovery-focused use cases. Also, to be clear, image analysis is not new in and of itself, and conceptually many theories are not new either, but the way that the data is being structured and the assumptions around the mainstream industry approaches are substantially different now.

Human-supervised training data and classic datasets are different approaches covering different problems. We shouldn't oversimplify them as being all the same—a spreadsheet is materially different from the assumptions and structures of data discussed in this book. We also shouldn't contrast them as competing; supervised training data is unlikely to help in discovery-focused use cases, so both approaches are still needed.

Discovery

Datasets classically have been about the discovery of new insights and useful heuristics. The starting point is often text, tabular, and time series data. This data is usually used as a form of discovery, such as for recommender systems (Netflix's suggested movies), anomaly detection, or car reconditioning costs.

In classic dataset techniques, there is no form of human "supervision." Humans may be involved in defining preprocessing steps, but they are not actively creating new data to the same degree as in annotation. Humans may also still be involved in feature engineering, or in a deep learning context (but still classic discovery focused) there may not even be feature engineering. Overall, the data is fixed, a mathematical algorithmic goal is defined, and the algorithm goes to work.

Feature engineering is a practice of selecting specific columns (or slices) that are more relevant, for example the odometer column for a vehicle reconditioning cost predictor.

With the new training data techniques, humans are supervising the data. We already know what the correct answer is, and the goal is to essentially copy that understanding, "Monkey See, Monkey Do."

This direct control makes the new techniques applicable to a variety of new use cases, especially in the realm of "dense" or "unstructured" data, such as video, images, and audio. Instead of the data being fixed, we can even control the generation of new data, such as taking new images.

General Concepts

When you are creating, updating, and maintaining training data, there are a number of concepts that come into play. While we cover many specifics earlier in the book, I want to take a moment to reflect on more general concepts that carry throughout multiple areas of training data:

- Data relevancy
- The need for both qualitative and quantitative evaluations
- Iteration
- Transfer learning
- Bias
- Metadata

Data Relevancy

Continuing with the theme of validation of existing data, how do we know if our data is actually relevant to the real world? For example, we could get a perfect 100% score on a test set, but if the test set itself isn't relevant to the real-world data, then we are in trouble!

There is currently no known direct test that can define if the data is guaranteed to be relevant to the production data—only time will truly tell. This is similar to the way a traditional system can be loaded and tested for x number of users, but the true usage pattern will always be different. Of course, we can do our best to plan for the expected use and design the system accordingly. Like in classic systems, tests and other tools can help, but you still need a good design in the first place.

Overall system design

There are usually many choices for how to design a data collection system in machine learning. For example, for a grocery store, a system could be positioned over an existing checkout register, such as to prevent theft, or aid in faster checkout, as shown in Figure 6-5.

Or, a system could be designed to replace the checkout entirely, such as placing many cameras throughout and tracking shopper actions. There is no right or wrong answer here—the primary thing to be aware of is that the training data is tied to the context it's created in. In later sections we will cover system design in much more depth.

Figure 6-5. Left, example of integration with an existing system; right, example of new process

Raw data collection

Beyond the high-level system design perspective, the collection and storage of raw data is generally beyond the scope of this book. This is, in part, because of the vast array of options available for raw data. Raw data can come from real-life sensors (video, audio, images), or it can be screenshots, PDF scans, etc. Virtually anything that can be represented in a human-parsable form can be used for raw data.

Need for Both Qualitative and Quantitative Evaluations

Both qualitative and quantitative types of evaluation are important. You'll want to apply the appropriate weight to quantitative methods. Quantitative metrics are tools, not overall solutions, and can't provide the comfort of knowing if the system is actually going to work. Often, these tools are "overweighted," and looked at as the oracle of system success, so you may need to decrease the weight of how significant these metrics are when viewed by you and your team.

Metrics-oriented tools also place more emphasis back on the data science team and further away from the experts who can actually look at the results one sample at a time—qualitatively, not just statistically. Most real-world systems don't work as well as is often believed by those who are overly guided by metrics. Metrics are needed and important. Just keep separate the also important, and always needed, human qualitative ability.

Iterations

In traditional programming, we iterate on the design of functions, features, and systems.

In training data, there is a similar form of iteration practiced using all of the concepts we are discussing herein. The models themselves are also iterative; for example, they may be retrained at a predetermined frequency, such as daily.

Two of the biggest competing approaches here are the "fire and forget" and the "continual retrain." In some cases, it may be impractical to retrain, and so a single, final model is created.

Prioritization: What to Label

As part of dataset construction, we know we need to create a smaller set from a larger set of raw data—but how? This is the concern of "What to Label." Generally, the goal of these approaches is to find "interesting" data. For example, if I have thousands of images that are similar, but three are obviously different, I may get the most bang for my buck if I start with the three most different ones. This is because those outliers will benefit the most from human judgments.

Transfer Learning's Relation to Datasets (Fine-Tuning)

As a brief primer, the idea of transfer learning is to start off from an existing knowledge base before training a new model. Transfer learning is used to speed up training new models. While the specifics of the implementation details of transfer learning are outside the scope of this book, I wish to draw attention to how some of the training data and dataset choices relate to and impact fine-tuning.

You may have heard that popular datasets used in transfer learning or pretrained models contain biases that are hard to remove. There are theories about the effects of those datasets on the researcher, including that the biases leak into the design of the models. This is often true to a degree, but is usually not particularly relevant to real-world supervised AI applications.

To get a better intuition, let's consider the context of pre-training for image detection tasks. A large portion of the pre-trained knowledge is things like detecting edges, basic shapes, or sometimes common objects. In this context these base level features are more an artifact of the training process than true "training data biases." In other words, they are features whose inclusion is either routine or their presence is not negative in a meaningful way.

A critical distinction here, relative to the earlier discussed concerns around historical datasets, is the original dataset is just a starting point, from which the real dataset will be fitted. In other words, if there's a problem in the real dataset, then yes the model will still be trying to learn that. But if there's a problem in the original set not present in the real set, the chances of it remaining during the fine-tuning or transfer learning processes are much lower.

To test this for your specific case, train both with transfer learning and without, and compare. Usually, these comparisons show that transfer learning is faster, but the actual end result, the end state of the model after the training process, is similar. You can do this as a check, and then for ongoing iterations you will know if you can likely use transfer learning "safely" with less bias concern.

The concern can be stated like this: "Because we are indirectly using training data from that prior model training, if there was undesirable bias in that model, it may carry over to our new case. There is a dependency on that prior training data set." However, we must remember the intent of transfer learning: a way to reach an optimal point faster. In practice, this means it's usually about getting general "lower-level" concepts to exist in the model (think edges, basic shapes, basic parts of speech, etc.). Moreover, most of the things we could consider to be human-level bias are overwritten on the first few training passes when the new labels are used as the source of error "loss."

As an example, consider that you can train on images with meanings totally different from the original set. For example, a model pre-trained on Common Objects in Context (COCO) (*https://oreil.ly/iMDel*) is a large-scale object detection, segmentation, and captioning dataset) has no concept of a product "SKU 6124 Cereal box." A commonly mistaken assumption is that using transfer learning is automatically better than not using transfer learning. Actually, you can achieve just as good results without transfer learning, as with it. For real-world applications, transfer learning is usually more of a compute optimization than a model performance or bias concern.

Technically, a model should be able to converge to the same point regardless of the starting weights from transfer learning. For context, the pre-trained weights represent the learned representation (higher dimensionality) of the original dataset. I used the terms "weights" and "data" interchangeably here to focus on the training data comparison, but in data science practice they are naturally different. There may be cases where there is not enough data to converge or overwrite the original weights completely, and data from the original set remains, e.g., in the form of unchanged pre-trained weights. If the original dataset contains useful data to the new task (data not in the new set) this remaining data may even be desired. The joint dataset (original plus starting weights) may now still get better performance than either set individually. Having a true joint dataset is less likely with larger new datasets, in which case more or all of the transferred set's weights get "overwritten." It also can depend on the context of the pre-trained set. If the intent was to make use of semantic information, that is a different use case than using it for "base" concept detection like edges, as mentioned earlier.

While more depth in that topic is beyond the scope of this book, essentially, as long as you are using transfer learning as an optimization (e.g., to save time training) and still have your own novel data over top of it, then it is unlikely that bias from the original set will be a material issue. Potential issues become even less noticeable as the ratio of the new dataset grows relative to the pre-trained set. On the other side of the coin, if for your case, the original dataset (through transfer learning) is remaining or affecting the final distribution of the data dramatically, then you should treat the data of the pre-trained model as part of your overall current dataset and inspect it closely for undesirable bias.

So to recap, there are two things to remember about transfer learning:

- It is not a replacement for having your own domain-specific (new) datasets.
- It is usually a compute or algorithmic optimization more than bias concern. However, if substantial parts of original dataset or weights are kept, and the weights are beyond basic internal concepts like edges and shapes, then it is a bias concern.[5]

Again, to be clear, if there is unwanted bias in the training set, and that set is directly used, that is bad. If a transfer learning set is used just to speed up the initial weighting setting of your dataset, then the risk is a lot lower. For any cases where you suspect bias to be an issue, train with and without the transfer learning set to compare and inspect the original set for undesirable bias.

This is an evolving and controversial area, and the above is meant only to provide a high-level overview as relevant to training data. Please note this is all in the supervised (human reviewing the data) context.

Per-Sample Judgment Calls

Ultimately, a human will supervise each sample, generally one sample at a time. In all of this, we must not forget that the decisions each person makes have a real impact on the final result. There are no easy solutions here when it comes to difficult judgment calls. There are tools available, however, such as taking averages of multiple opinions, or requiring examinations.

Often people, including experts, simply have different opinions. To some extent, these unique judgments can be thought of as a new form of intellectual property. Imagine an oven with a camera. A chef who has a signature dish could supervise a training dataset that, in a sense, reflects that chef's unique taste. This is a light introduction

---

5 Note that theoretically, it can always carry over unwanted bias. But in practice it's often not that relevant, since the most "important" weights are often overwritten by the new dataset.

to the concept that the line between system and user content becomes blurred with training data in a way that's still developing.

Ethical and Privacy Considerations

First, it's worth considering that some forms of supervised data are actually relatively free of bias. For example, it's hard for me to imagine any immediate ethical or privacy concerns from our strawberry picking dataset. There are, however, very real and very serious ethical concerns in certain contexts. This is not an ethics book, though, and there are already some relatively extensive books on the effects of bias in automation more generally. However, I will touch on some of the most immediate and practical concerns.

Bias

The term "bias" is overloaded; it has many many meanings. Here, I will disambiguate just a few; there is technical bias, imbalanced classes bias, desired human bias, and undesirable human bias.

First, even within technical bias, there are multiple meanings for the term. In ML modeling, bias can mean the fixed value that is added to the variable part of a calculation. For example, if you wanted the model to return 3 when the sum of the weights is 0, you can add a bias of 3.[6] It can also mean the estimator bias as measured by distance from the mean of the real distribution (which, for real-world applications, is never fully known). These measures may be useful to data scientists and researchers, but it's not the bias that we are concerned about, which is human bias.

Imbalanced classes are when one label has more samples than another. This is easy to state, but how to fix it, in practice, is undefined. To understand why, consider the following example: imagine that we are designing a threat detection system for an airport scanner, as shown in Figure 6-6.

We may have classes like "forearm" and "threat." How many forearm examples do we need? How many threat examples do we need? Well, the forearm, in this context, has a very low variability, meaning that we likely only need a small sample set to build a great model. However, the possible variations in threat placement, and the purposeful efforts to obscure it, mean we likely need many more examples.

---

6 Example inspired by article by Farhad Malik, "Neural Networks Bias and Weights: Understanding the Two Most Important Components" (*https://oreil.ly/MdbQu*), *FinTechExplained—Medium*, May 18, 2019.

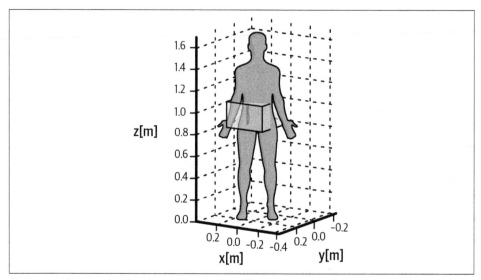

Figure 6-6. Example of raw scan from 3D millimeter wave scanner

At first glance, in that context, then, "forearm" may be out of balance to "threat," but it's actually desirable, because "threat" as a concept is a harder problem than "forearm." Maybe it needs 10× as much data, or maybe 100×—the absolute count of samples doesn't matter as much as how much data is needed to be performant *on that class*. This is the imbalanced classes problem. It may take 10× as much data for label A to be "balanced" with label B. Another way to approach this is to subdivide "threat" into smaller categories.

This leads us to a slightly more subtle problem. We have been assuming, in our example, that all instances of a "threat" are equal. But what if the training data isn't representative of the real-life data? This is technically a different concept, the data distribution.

The thing is that correcting "obvious" bias has a variety of technical solutions, such as hard negative mining, but correcting the relationship of the data to the real world does not. This is similar to how, for example, a classic program can pass hundreds of "unit tests," but still utterly fail to meet the end user's needs.

The model needs to be "biased" toward detecting what you want it to detect. So, keep in mind that from that perspective, you are trying to make the model "biased" toward understanding your worldview of the data.

Bias Is Hard to Escape

Imagine this scenario:

- A dataset is created in a given month
- To maintain freshness, only "new" data from the last six months is used
- To optimize the model, a sampling of the output and errors is reviewed and corrected by humans

What this really means though, is that every "new" example is recycled into the model. In other words, say that a prediction and subsequent correction happens on day one. How long can we use this? We presume this "fresh" correction is valid for six months. But is it really?

Well, the reality is that even if it's "correct," after six months, its basis is a model that is now old. This means that even if we retrain the model using only data corrected within the last six months, there is still bias from the "old" model creeping in.

This is an incredibly hard problem to model. I'm not aware of a scientific solution here. This is more of a process thing to be aware of—that decisions made today can be difficult to completely roll back tomorrow.

An analogy in coding may be system architecture. A function may be relatively easy to "correct," but it's harder to realize if that function should exist at all. As a practical matter, an engineer working to correct an existing function will likely start with that function, so even the corrected function will contain the "spirit" of the old one, even if literally every character is changed.

Besides the literal data and instances, a further example here is the label templates. If the assumption is to always use the existing predictions, it may be hard to recognize if the templates are actually relevant any more.

Metadata

Imagine you spent thousands of hours (and potentially hundreds of thousands of dollars) to create multiple datasets, only to realize that it wasn't clear what assumptions were present when it was created. The implication of this is the data could be less useful than previously expected, entirely useless, or even worse, it could have excess negative effects due to unexpected results when used outside the expected context.

There are many reasons, some described following, that a dataset that is technically complete can become largely unusable. A surprisingly common problem is losing information on how a set was constructed. Imagine, for example, that a data supervisor has a question about a project and sends it over a channel like email or chat. The problem, such as "how do we handle a case when such and such happens?" is resolved and life goes on. However, as time progresses that knowledge is often lost. If that issue arises again, the process of querying the colleague and waiting for a response has to happen again or in some cases the raw records or colleague may not be available. Therefore it's best to maintain as much metadata as reasonably possible for your context, e.g., by including it in definitions in your system of record.

Metadata of a Dataset

Data about the dataset that is not directly used by the model. For example, when the set was created, who created it. Don't call annotations "metadata," because annotations are a primary component of the overall training data structure. Consider that if we called annotations metadata, then metadata becomes meta-metadata which doesn't make sense.

A small note is that some people refer to any type of annotation as metadata. If we take an old textbook definition, this is reasonable. In practice, annotations and schemas are thought of as separate things from generic metadata.

Preventing Lost Metadata

There are many examples of metadata that is commonly "lost" from sets:

- What was the original purpose of the set?
- Who were the humans who created it?
- What was the context in which the humans created it?
- When was the data captured? When was it supervised?
- What were the sensor types, and other data specifications?
- Was the element made by machine originally, or by a human? What assist methods, if any, were used?
- Was the element reviewed by multiple humans?
- What other options (e.g., per attribute group) were presented?
- Was the templating schema changed during set construction? If so, when?
- How representative is the set versus the "original" data?
- When was the set constructed? For example, the raw data may have a timestamp, but that is likely to be different from when a human looked at it (and this is likely to be different *per sample*).
- What guides were shown to supervisors? Were those guides modified, and if so, when?

- What labels are in this set and what are the distributions of the labels?
- What is the schema of relating the raw data to the annotations? For example, sometimes annotations are stored at rest as a filename like *00001.png*. But that assumes that it's in the folder "xyz." So if for some reason that changes, or it isn't recorded somewhere, it can be unclear which annotations belong to which samples.
- Is this only "completed" data? Does the absence of an annotation here mean the concept doesn't exist?

Metadata loss can be avoided by capturing as much information as reasonably possible during the creation process. In general, using a professional annotation software will help with this.

Train/Val/Test Is the Cherry on Top

When creating a model, a common practice is to split an original set into three subsets; this is called *Train/Val/Test*. The idea is to have a set that the model is trained on, a second set that's withheld from training to validate on, and a third set that's held in reserve until work is complete, to do a one-time final test on. It's relatively trivial to sample an existing set and split it.

However, where did the original "set" come from?! This is the concern of the training data context: constructing the original, updated, ongoing sets themselves.

There are a variety of considerations around splitting these subsets, such as randomly selecting samples to avoid bias, but that's more the concern of data science than training data.

Usually, there is more raw data than can be annotated. So, part of the basic selection process is choosing which of the raw samples will be annotated. There are normally multiple original sets. In fact, many projects quickly grow to have hundreds of these original sets. More generally, this is also a concern of the overall organization and structure of the datasets, including selecting which samples are to be included in which sets.

Sample Creation

Let's explore, from the ground up, how to create a single sample of training data. This will help build understanding of the core mechanics of what the literal supervision looks like.

The system will have a deep learning model that has been trained on a set of training data. This training data consists of two main components:

- Raw images (or video)
- Labels

Here, we will discuss a few different approaches.

Simple Schema for a Strawberry Picking System

Let's imagine we are working on a strawberry picking system. We need to know what a strawberry is, where it is, and how ripe it is. Let's introduce a few new terms to help us work more efficiently:

Label

A label, also called a *class*,[7] represents the highest level of meaning. For example, a label can be "strawberry" or "leaf." For those technical folks, you can think of it as a table in a database. This is usually attached to a specific *annotation* (instance).[8]

Annotation (instance)

A single example is shown in Figure 6-7. It is connected to a label to define *what* it is. Labels also usually contain positional or spatial information defining *where* something is. Continuing with the technical example, this is like a row in a database.

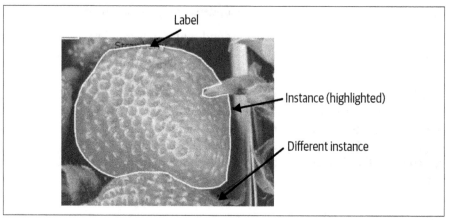

Figure 6-7. Labeled and not labeled instances

---

7 Other names include: label template, annotation name, class.

8 Usually by reference ID only.

Attributes

> An instance may have many attributes. Attributes are things that are unique to a specific instance. Usually, attributes represent selectable properties of the object itself, not its spatial location.

This choice will affect the speed of supervision. A single instance may have many unique attributes; for example, in addition to ripeness, there may be disease identification, produce quality grading, etc. Figure 6-8 depicts an example.

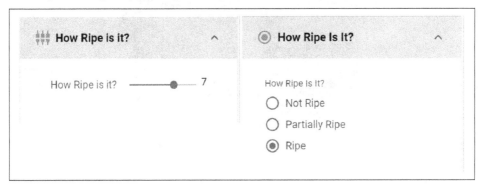

Figure 6-8. Example of UI showing choices for attributes

Imagine you only want the system to pick strawberries of a certain ripeness. You may represent the ripeness as a slider, or you may also have a multiple choice option about ripeness, as shown in Figure 6-8. From the database perspective, this is somewhat akin to a column.

Geometric Representations

You'll also make choices about what type of geometric representation to use. Your choices here are as much part of training data as doing the literal supervision.

Geometric shapes can be used to represent objects. For example, we can represent the strawberries shown in Figure 6-9 as boxes, polygons, and other options. In other chapters, I discuss these trade-offs.

From a system design perspective, there can be choices around what type of raw data to use, such as image, audio, text, or video. Sometimes, you may even combine multiple modalities together.

Figure 6-9. Annotation example with box and polygon geometric shapes

Angle, size, and other attributes may also apply here.

Spatial locations are often singular in nature. Now, while an object is usually only in one place at a given moment in time,[9] in some cases, a single frame may have multiple spatial annotations. Imagine a beam obstructing the view of a car: to accurately label the area of the car, we may need to use two or more closed polygons related to the same annotation.

Binary Classification

A basic approach is to detect the difference between something existing and not existing. One way to do that is called binary classification. This is akin to asking "Does the traffic light exist in the photo or not?"

As an example, two of the images in the set may look like Figure 6-10:

---

9 That is, in one reference frame. Multimodal annotations can be represented as a set of related instances, or spatial locations for a given reference frame, such as camera ID x.

Figure 6-10. Left, an image in a dataset showing a traffic light with trees in the background; right, an image in a dataset showing trees

To supervise Example 1, we need only two things:

- To capture the relation to the file itself: e.g., that the filename is *"sensor_front_2020_10_10_01_000."* This is the "link" to the raw pixels. Eventually, this file will be read, and the values converted into tensors, e.g., position 0,0, and RGB values will be assigned.

- To declare what it is in a meaningful way to us, for example: "Traffic_light" or "1." This is similar to saying filename *"sensor_front_2020_10_10_01_000"* has a "Traffic_light" existing in it.

And for Example 2, we could declare it as:

- "sensor_front_2020_10_10_01_001"
- "No" or "0"

In practice, there will usually be more sets of images, not just two.

Let's Manually Create Our First Set

You can do this with a pen and paper or white board. First, draw a big box and give it a title, such as "My First Set." Then draw a smaller box, and put a sketch of a traffic light and the number 1 inside it. Repeat that two more times, drawing an image without a traffic light and a 0, as shown in Figure 6-11.

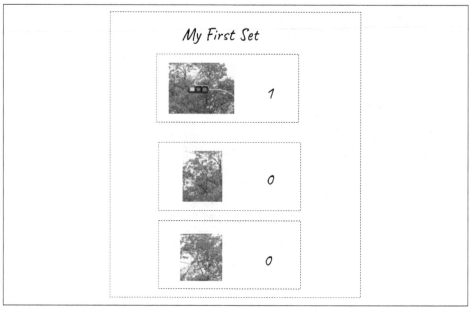

Figure 6-11. Visual example of a simple training data set

This is the core mapping idea. It can be done using pen and paper and can also, of course, be done in code. Realistically, we will need proper tools to create production training data, but from a conceptual standpoint, each methodology is equally correct.

For example, consider this Python code. Here, we create a list and populate it with lists where the 0th index is the file path and the 1st index is the ID. The completed result (assuming the literal bytes of the .jpgs are available in the same folder) is a set of training data:[10]

```
Training_Data  = [
['tmp/sensor_front_2020_10_10_01_000.jpg',     1],
['tmp/sensor...001.jpg',                             0],
['tmp/sensor...002.jpg',                             0]]
```

This is missing a label map (what does 0 mean?). We can represent this simple dictionary as:

```
Label_map = {
        1 :  "Traffic light",
        0 :  "No Traffic light"
}
```

---

10 Sharp-eyed readers may notice this becomes a matrix once completed. The matrix shape has no significance here, and it's generally best to think of these as a set. However, Python sets introduce quirks also not relevant here—so I use a list.

Congrats! You have just created a training dataset, from scratch, with no tooling, and minimal effort! With a little wrangling, this would be a completely valid starting point for a basic classification algorithm.

You can also see how, by simply adding more items to the list, you can increase the training dataset.

While "real" sets are typically larger, and typically have more complex annotations, this is a great start!

Let's unpack the human algorithm we are using. In sequence, we would do the following:

1. Look at picture.
2. Draw on knowledge of traffic lights.
3. Map our knowledge of traffic lights to sparse value, e.g., "Traffic light is present."

While it's obvious to us, it's not so obvious to a computer. For example, prior approaches in this space, such as Histograms of Oriented Gradients (HOGs) and other edge detection mechanisms, have no real understanding of "traffic light" and "not traffic light."

For the sake of this example, I made a few assumptions. For instance, I imagined the traffic lights were pre-cropped by some process, the angle of the traffic lights was presumed to be correct, etc. In practice, it's common to use a spatial shape, such as a box or a cuboid so that the location can be determined by the model (e.g., object detection). Also, there may be many other states or pre-processing pipelines.

Of course, in production, the processes become more complex, but this example demonstrates the aspect of this process is classification and mapping our understanding to the computer. There are multitudes of more complex forms of doing this, but this is the heart of the idea. If you understand this, you are already well ahead of the game.

Ultimately, other approaches typically build on classification, or add "spatial" properties to samples. At the end of the day, even if some prior process runs, there's still a classification process being done.[11]

---

11 Algorithms usually predict a continuous range and then do a function (e.g., softmax) to convert to these category values.

Upgraded Classification

To expand from "it exists" to "what is it" we will need multiple classes. Table 6-2 shows an example of how this is laid out.

Table 6-2. Visual comparison of raw data and corresponding label

Sample #	Raw media	Label name	Integer ID
Sample 1	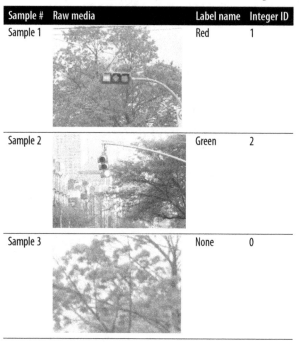	Red	1
Sample 2		Green	2
Sample 3		None	0

With regard to whether to use strings or integers, generally speaking, most actual training will use integer values. However, those integer values typically are only meaningful to us if attached to some kind of string label:

```
{ 0 : "None",
  1 : "Red",
  2 : "Green"}
```

Here, I will introduce the concept of the label map. While mapping of this type is common to all systems, these label maps can take on additional complexity. It's also worth remembering that in general, the 'label' means nothing to the system, but it's mapping the ID to the raw data. If these values are wrong, it could cause a critical failure. Worse—imagine if the testing relied on the same map!

That's why, when possible, it's good to "print output"; that way you can visually inspect whether the label matches the desired ID. As a small example, a test case may assert on a known ID matching string.

Where Is the Traffic Light?

Continuing with the traffic light example, the problem with the preceding approach is that we don't know *where* the traffic light is. There is a common concept within ML modeling called an *objectness score* that we touched on earlier in the book. And there are other, more complex ways to identify location. From a training data perspective, as long as we have a bounding box present we have identified the *where* (the spatial location), the implementation of the algorithm to learn that is up to data science.

Maintenance

Now that we have covered the basics of how a single sample is created, and introduced some key terms, let's zoom back out to a high-level look at the process. In Chapter 2, we covered basics like setting up your training data software, schema, and tasks. But what do ongoing maintenance actions really look like?

Actions

These are actions that can be taken, in general, after some form of information has been returned from the model training process.

Increase schema depth to improve performance

One of the most common approaches to improve performance is to increase the depth of the schema. One example is to divide and conquer the label classes, especially poorly performing classes. Essentially, this is to both identify and improve the specific classes that are the weakest. To borrow from our earlier example, let's use the label "Traffic Light." When there is mixed performance, it may be unclear which examples are needed to improve performance.

When reviewing the results, we notice that "Green" seems to show up more often in the failure cases. One option is to try to add more green ones to the general traffic light set. A better option is to "split" the class "Traffic Light" into "Red" and "Green." That way we can very clearly see which is performing better. We can repeat this until the desired performance is reached. We do this by again, splitting between large and small as shown in Figure 6-12. There are a few intricacies and approaches to the implementation of this, but they generally revolve around this idea. The key idea is that at each "split," it becomes clear which way to go next in terms of performance.

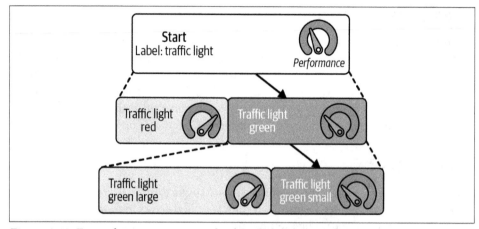

Figure 6-12. Example improvement path of single label expanding into multiple, more specific, attributes based on performance needs

Better align the spatial type to the raw data

Imagine you started with choosing segmentation. Then, you realize the model is not training as desired. You may be able to simply switch to an "easier" task like object detection, or even full image classification.

Alternatively, perhaps object detection is yielding a bunch of overlapping boxes which aren't helpful, and you need to switch to segmentation to accurately capture the meaning, as shown in Figure 6-13.

Figure 6-13. Boxes failing to provide useful info; example of moving to segmentation to get better spatial results

In the left side of the diagram, we start with boxes, which results in the overlapping detections, as shown on the left. By moving to segmentation, we can get more clear segments, as shown on the right. While it may appear clear which methods are less ideal for certain cases, the optimal method is often less clear.

Create more tasks

Annotating more data for better performance has almost become a cliche already. This orientation is often combined with the other approaches, for example, dividing the labels or changing the spatial type, and then supervising more. The primary consideration here is whether more annotation will provide net lift.

Change the raw data

In some cases, it's possible to change the raw data sourcing or collection mechanisms. Examples include the following:

- Change the sensor type or angle.
- Change what part of the screen is being captured.

Net Lift

As we annotate a dataset, we usually want to do two things:

- Identify previously unknown cases or problem cases.
- Maximize the value of each net new annotation.

To illustrate the need for net lift, consider a raw, unlabeled dataset consisting of hearts, circles, and triangles, as shown in Figure 6-14.

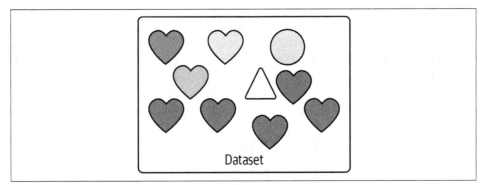

Figure 6-14. Example dataset with abstract shapes to represent data elements

If we annotate only hearts, then each heart we supervise provides minimal value. It's also a problem that we don't understand the complete picture because we did not annotate any circles or triangles. Instead, we should always try to annotate to get true *net lift*, which often means annotating samples different enough to provide value to the model (see Figure 6-15).

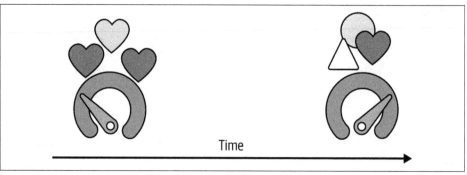

Figure 6-15. Example of better performance over time by annotating different shapes, not just slightly different hearts

There are a few ways we can test if we have seen enough of one sample type:

- Relative performance of that label type is high (as shown in "Labels and Attributes—What Is It?" on page 68).
- A process like similarity search has identified overly similar data elements (or missing elements).
- Humans have looked at the data and feel it's very similar or know that elements are missing.

Note that a model may need less data for one class than another—sometimes by multiple orders of magnitude. This is completely OK, and achieved in data science via methods like using multiple models or multiple "heads" of the same base model. It is not about having 1:1:1 ratios for every label, but rather labels that will well represent the difficulty of the problem. To extend and reframe the preceding example, if it really matters that we detect slightly different colors of the hearts, then we may need dramatically more heart examples than triangles or circles.

Levels of System Maturity of Training Data Operations

From early exploring, to proof of concept, through to production, common ways to identify the stage you are in are defined in Figure 6-16.

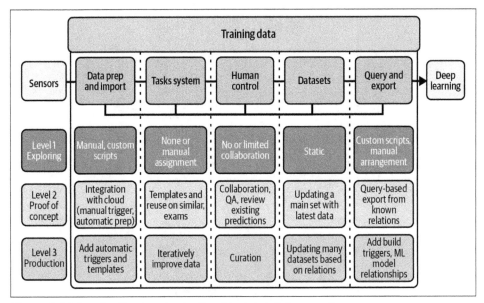

Figure 6-16. Levels of system maturity of training data operations

Applied Versus Research Sets

The needs and form of training data continue to rapidly evolve. When people think of datasets, popular research datasets, like MS Common Objects in Context (COCO) 2014 (*https://oreil.ly/iMDel*), ImageNet 2009 (*https://oreil.ly/cv2hP*), (both vision) and the General Language Understanding Evaluation (GLUE) 2018 (*https://oreil.ly/_1Ji-*)[12] come to mind. These sets are designed for research purposes, and by design, they evolve relatively slowly.

In general, it's the research that's designed to change around these sets, not the sets themselves. The sets are used as benchmarks for performance. For example, Figure 6-17 shows how the *same* set is used in five different years. This is useful for research comparisons. The core assumption is the sets are static. The key point here is the dataset is not changing, in order to compare algorithms; however, this is opposite what often happens in industry, where the algorithms change less and the data changes more.

---

12 These sets all have taken an incredible amount of work, and some have been substantially updated over time. Nothing in here is meant to take away from their contributions.

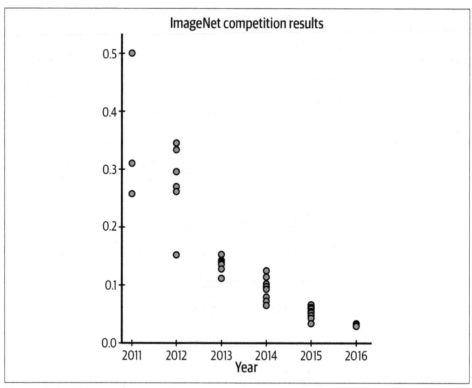

Figure 6-17. Showing the same dataset getting a lower error rate over time due to improvements in studied algorithms (Source: Wikipedia (https://oreil.ly/FjUH0))

In the context of a practical commercial product, things change. In general, the rule of thumb is that data more than 12 months old is unlikely to accurately represent the current state. Generally, the assumption for practical sets is that they are only static for the moment of literal training, but are otherwise always changing. Further, commercial needs are very different from research. Time available to create the sets, time to update, costs to create the sets, implications of mistakes, etc., are very different in the two contexts.

Training Data Management

We have touched on some of the maintenance concerns of the process. Training data management is concerned with the organization of humans doing the literal tasks and the organization of data throughout the entire operations cycle.

One of the central ideas behind training data management is maintaining the "meta" information around the training data, such as what assumptions were present during the creation of the data. In the context of continually improving and reusing data, for a large organization this is especially important. Imagine spending a significant budget on setting up pipelines, training people, and creating literal sets, only to "lose" that information because of improper handling.

Quality

Quality discussions can escalate quickly into religious-type arguments, with people mystically conjuring statistics, and others rallying hordes of superstitions to their side.

There are a few things to remember during this excitement:

- All datasets contain errors, and yet many ML programs still obtain acceptable and useful results. This shouldn't actually be a surprise, because it's similar to how all programs have some degree of bugs.
- Neither all lines of a code program, or a data program, exist in a vacuum. Each has a variety of cues, visible at various points, that act as checks and balances. If I see a loading icon stuck, I know something is wrong. As an annotator, if I see that none of the schema matches what I am seeing, I know something is wrong.
- If the margin of error is such that many people need to look at it to get the right answer, the schema is probably wrong.

The annotation quality is itself just one aspect of the overall quality, and it isn't improved just by "adding more" annotations. Sizing the dataset appropriately also impacts the overall quality of the model. This is another reason why using a custom dataset that is custom created to suit your needs is important. If the sample size is too large (or too small), the output will be negatively impacted. Too much extraneous data can skew the results and produce more errors from a statistical perspective, and too little data won't accurately depict the model.

Annotation quality is never fully solved solely by just doing more of it, duplicating samples to more people, etc. This is similar to how "more" education, especially repeating similar lessons, doesn't automatically make you a more effective worker.

Completed Tasks

Knowing which samples are completed at any given moment in time is surprisingly challenging. At a high level, this is partly because we are trusting other humans to identify what is completed and what is not. Second, because the schema often changes, the definition of complete also changes. For example, a file may have been completed relative to a schema that is no longer relevant.

The most minimal management need is to separate "complete" samples from "incomplete" samples. It matters because any sample trained on is considered to be valid to the network. So, for example, if a sample is included with no further labels, it will be considered background.

Why "complete" matters:

This can cause severe problems:

- It makes it hard to create a performant network/debug it.
- If this class of error makes it to a validation test set, the test set will be equally affected.

While this may seem trivial in small set sizes, in most significant sets:

- The data scientist(s) will never observe 100% of the samples, and sometimes will never observe a significant set of the samples, or even *any*! In the case of using transfer learning, e.g., pre-train on ImageNet, anyone who has ever used ImageNet has done this!
- In large set sizes, it is unreasonable to expect any single person to be able to review all of it.

To make an analogy to web design, this is the "test case passed, but important checkout button is invisible to the user and our sales stopped" case for training data.

In general, one should use software to manage this process and ensure that completed/not completed is well tracked. It's also worth considering that in many externally created datasets/exported datasets, this "completed" tracking gets lost, and all of the samples are assumed to be completed. Three common cases showing "completed" status changes are listed here in Table 6-3.

Table 6-3. Tracking of "completed" statuses

Tracking "completed" status context	Reasonable way to track "completed" work status
Human review	Use the "review loop" concept. This means the file status is not complete until the review loop is completed.
Schema changes	Tasks have a schema related to them. If the schema changes, then the "completed" status should be considered questionable.
Multimodal	Use compound files (or tags, groups of tasks) to group multimodal elements together, such that when a compound file is completed, it's completed as one thing.

Freshness

Certain aspects of training data "age well." What does age well:

- Transfer learning–type concepts/"lower-level" representations, likely on the order of decades (yet unproven)
- Tightly controlled scenarios (e.g., some types of document reading, medical scanning), likely on the order of years

What doesn't age as well:

- Novel scenarios /unconstrained real-world scenarios (e.g., self-driving cars)
- Anything that involves a sensor that's not in tightly controlled space., e.g., anything outside
- Specific "heads/endpoints" labels
- Step function changes in the distribution, e.g., sensor type changes, scenario changes (day/night), etc.

Different applications naturally have different freshness requirements. However, some models that appear to "last a long time" are doing so because they were relatively overbuilt to begin with.

Part of the "trick" with the whole freshness aspect is determining which samples to keep. If possible, it's good to test and compare a rolling window versus continuous aggregation.

Maintaining Set Metadata

In the rush to create training data, we often lose sight of key contextual information. This can make it very difficult to go back, especially as one develops many different sets. A training data set without context to a real distribution, i.e., the real world, is of little value.

Task Management

The reality is that with humans involved, there is inevitably some form of organization or "task" management system needed—even if it's "hidden" behind the scenes. Often, this ends up involving some form of people-management skills, which although important, is beyond the scope of this book. When we talk about tasks, we will generally remain focused on training data–specific concerns, such as tooling and common performance metrics.

Summary

Creating useful training data for machine learning systems requires careful planning, iterative refinement, and thoughtful management. We explored theories around how training data controls system behavior by encoding human knowledge as the ground truth goals for machine learning algorithms:

Theories
: A system is only as useful as its schema. Who supervises the data matters. Intentionally chosen data works best. Training data is like code. Focus on automation over discovery. Human supervision differs from classic datasets.

Concepts
: Data relevancy, iterations, transfer learning, bias, metadata.

Sample creation
: Defining schemas, labels, annotations, attributes, and spatial types. Binary and multi-class classification examples.

Maintenance
: Changing labels, attributes, spatial types, raw data. Considering net lift from new annotations. Planning system maturity levels.

Management
: Tracking quality, task completion, freshness, metadata, and evolving schema.

Key takeaways:

- Alignment between schema, raw data, and annotations is critical.
- Managing the training data lifecycle complexity is key, especially for large, real-world systems.
- Training data controls system behavior by defining ground truth goals. Like code, it encodes human knowledge for machines.

You learned about how to increase schema depth to improve performance, and other actions like better aligning spatial types. We went over maintenance concepts like focusing on the net lift of each new annotation.

I introduced the major training data parts of MLOps and a path for planning system maturity. I introduced the concept of training data management: the organization of training data and people.

CHAPTER 7

AI Transformation and Use Cases

Introduction

This is the start of the AI transformation era. The most successful companies will be the ones who embed AI most deeply, who infuse AI broadly throughout the entire org, and push responsibility to use, train, and manage AI down to the lowest capable levels of the firm.

In this chapter I will cover two major topics:

- AI transformation and leadership
- Use case discovery, rubrics, and unlocking your own experts

This chapter also has extensive guidance on how to identify high-impact areas to apply AI. Examples include evaluation rubrics you can use to test your ideas, example questions to ask, and common missteps to avoid. If you want to dive straight into implementation details, the use case section may be most valuable to you. Moving to implementation, it's important to set a talent vision, which is something we'll go over. "The New "Crowd Sourcing": Your Own Experts" on page 229 will empower you to achieve your use cases in a cost-effective manner by repurposing existing work.

After reading this chapter, you will have a strong grasp of how to:

- Start your AI transformation at your work today.
- Evaluate AI use cases for your business.
- Define training data knowledge needs for yourself and your organization.
- Understand how to adopt modern tools and start aligning your team, peers, and staff with the goals of working with training data.

When you have the mindset, the leader, the use case, and the talent vision, you can use modern training data tools to make it a reality. I'll conclude this chapter by grounding these concepts into tooling. Training data tools have advanced considerably in the past few years, and there's a lot to learn.

The age of AI is here. This chapter will provide you with the mindset, leadership skills, use cases, and talent strategy needed to transform your company and emerge as an AI leader.

AI Transformation

AI transformation is the answer for business leaders who want to transform their company to take advantage of AI first. It can start today with you. So far, we have covered the basics of training data and taken deep dives into specific areas like automation. Now, we zoom out to see the forest. How do you actually get started with training data at your company?

In this chapter, I will share five key steps, depicted in Figure 7-1. We'll start with mindset and leadership, then move on to the concrete problem definition, and conclude with the two key steps to succeeding: annotation talent and training data tools. Please consider the plan merely a starting point to be adapted to your needs.

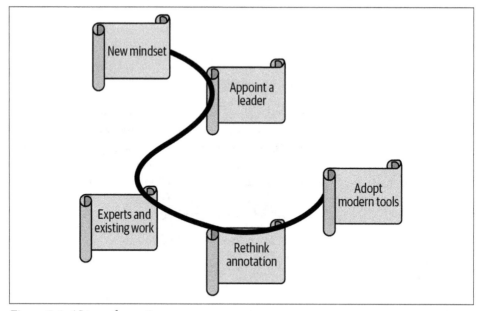

Figure 7-1. AI transformation map

To get started fast at adopting modern AI in your company, here are five key things to act on:

- Create new meaning instead of analyzing history—the creative alignment revolution.
- Appoint an AI data leader to lead the charge.
- Focus on use cases involving experts, using existing high-volume work.
- Rethink annotation talent—your best aligners are already working for you!
- Adopt new training data tools, like AI datastores, UI/UXs, and workflows.

For the last decade or so, the big wins with AI have been with classical cases, as discussed in earlier chapters.

Now the biggest commercial opportunities are with supervised learning, including the human alignment of generative AI systems like large language models. This supervision is about unstructured data that requires annotation (alignment).

Early work was often a case of putting together an "annotation project." This meant throwing data over a fence to some team and hoping for the best. This is like ordering fast food. Sure, it will help with immediate hunger, but it's not a healthy long-term solution.

True AI transformation, like eating healthy, takes work. It's a mindset shift, from seeing annotation as a one-off project, to seeing your day-to-day work as annotation.

Seeing Your Day-to-Day Work as Annotation

To frame this, consider this statement with regard to your company:

> All of your company's day-to-day work can be thought of as annotation.

That's right. Every action that the majority of your employees take, every day, is, both literally and figuratively, annotation. The real question here is, how can we shift those actions from being *lost*, every day, to being captured in a way that can be repeated, to the same quality degree, automatically? This mindset shift is reflected in Figure 7-2.

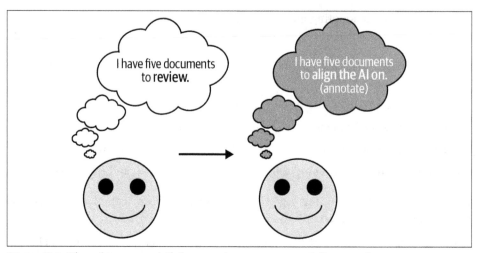

Figure 7-2. Thought process shift from review to annotate (alignment)

To invert this, every moment an employee does something that's not being captured as an annotation, that's productive work lost. The greater the percentage of that work that can be captured through annotation, the greater your productivity is. A further comparison is shown in Table 7-1.

Table 7-1. Comparison of classic working style versus AI training–centric data capture

Before	After
All day-to-day work is "one-offs."	Day-to-day work is handling the exceptions not yet captured in annotations.
Training is for humans only.	Training is for machines too.
If it's not in the computer, it doesn't exist.	If it's not in training data, it doesn't exist.

A rough analogy here is that in the movement from analog to digital, we realized that if it didn't exist in a digital form, then it didn't "exist" (whether that was true in reality or not). Now in the same way, if it doesn't exist in data, in this case AI training data, it might as well not exist either, since it won't help your company improve its productivity.

Let's put some structure to this theory.

There are two major types of AI transformation:

- Inspiring all relevant aspects of operations to give consideration to AI and establishing new reporting units
- Inspiring a training data–first mindset and reorganizing reporting relations

At a classic company, it's usually about building large-scale systems and adopting AI tools.

At an AI product company, it's about keeping up with the latest trends, updates, and AI tooling. Usually, an AI product company is more about technical infrastructure–level tools, whereas a classic company may use tools produced by said AI companies.

The Creative Revolution of Data-centric AI

Data-centric AI can be thought of as focusing on training data in addition to, and as being even more important than, data science modeling. But this definition does not really do it justice. Instead, consider that data-centric AI is more about creating new data to solve problems.

The creative revolution is a mindset of using human guidance to create new data points to solve problems. Opening up to the huge magnitude of this potential will help prime you and your team to define the best use cases. Next is appointing someone to lead the charge—a Director of Training Data. This position will be new in most companies. Given that a major portion of AI costs is made up by human labor in producing training data, this position naturally needs someone to account for it.

You Can Create New Data

In the data-centric mindset, the critical realization is that you can do one of two things:

- Use or add new data
  - New connections to existing files and data
  - New sensors, new cameras, new ways to capture data, etc.
- Add new human knowledge
  - New alignment and annotation

For example, for a self-driving car case, if you want to detect people getting cut off, you can create the meaning of "Getting Cut Off" as shown in Figure 7-3.

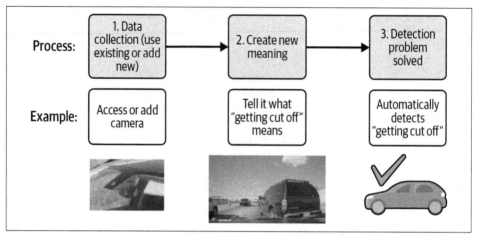

Figure 7-3. First example of creating new data for Data-centric AI

Or if you want to automatically detect what a "foul" means, you can create that too, as shown in Figure 7-4.

Figure 7-4. Second example of creating new data

You Can Change What Data You Collect

This may be obvious, but let's consider how different the following two scenarios are. Classically, in data science, you could *not* change the data you collected. For example, if you were collecting sales data, the sales history was just that—history. Small specifics aside, the sales were whatever they were. You couldn't invent new sales, or really change the data.

With training data, you can literally collect new data. You can place new cameras. Launch more satellites. Install more medical imaging devices. Add more microphones. Change the frequency of collection. Increase the quality, e.g., the resolution of the camera.

By focusing on the data that you can control, you can directly improve the performance. Better camera angle? Better AI. More cameras? Better AI. More...? I think you get the picture.

You Can Change the Meaning of the Data

Back to our sales example. A sale is a sale. There's little value in trying to expand a row in a spreadsheet to mean something more than it is.

With training data, you literally create meaning that was in no way, shape, or form there before. You look at a piece of media, like an image, for which the computer had no meaningful structure of interpretation before—and now you can literally say, "this is a human," "this is a diet soda," "this is a lane line." This act of annotation maps your knowledge into the computer.

The limit is only your creativity. You can say "this human is sad," or "this diet soda has a dent in it." You craft and mold and modify it to your needs. This quality of having infinite degrees of freedom is what makes it so powerful.

You Can Create!

So next time someone says to you that "data-centric AI is a way to get better model performance," you will know it's so much more than that! It means you can change the data you collect, and the meaning of that data. It means you can encode your understanding of a problem in a whole new way. It means you can define the solution even if no solution existed prior. You can *create*!

Think Step Function Improvement for Major Projects

Some of the efforts described next are major projects requiring many people, resources, and years of effort to complete. Think of them as conceptual thought starters.

We'll start by considering retail shopping. In that context, with these new methods, we can supervise machines to tell them what people look like and what groceries look like.

This unlocks all-new use cases, like entirely replacing the cashier. This is *not* 5% better pricing. This is a fundamental shift in how we shop for groceries and design stores. Further comparisons are shown in Table 7-2.

Table 7-2. Conceptual before and after capturing supervision for major projects

Concept	Before	After
Shopping	Every time I shop, every time a cashier checks me out, that work is lost.	Annotation of people shopping.
Driving	Every time I drive, the effort put in, the work, is "lost."	Annotation of driving. Professional annotators annotate common scenes. My driving is captured to aid in this effort.
Document review (imagine loans, requests, etc.)	Every document I review, the work is lost.	Annotation of document. Work is captured to reduce similar work in the future.

The key insight, which bears repeating: anything you can annotate can be repeated.

Build Your AI Data to Secure Your AI Present and Future

In the ideal case, every dollar invested in aligning AI has the potential to add value to every similar employee's role, or even generate all-new revenue channels. Keep in mind that while this is true, and possible, there are many challenges and risks with AI. Most AI projects require human engagement and oversight, and it's possible for projects to fail completely and never reach a minimum level of performance required to deploy and get value back. The more your team is able to learn the concepts in this book, the more they will be able to contribute to production success and a real return on your AI investment. By building your AI data transformation, you are securing your AI present and future. It will completely transform your existing projects and help protect you from competitors in the future.

Appoint a Leader: The Director of AI Data

All revolutions need leaders, someone to preach the new message. To rally the troops. To reassure doubts. And the leader must have a team. In this section, I'll lay out best practices and common job roles and discuss how they all come together to form an optimal team structure to support your training data revolution.

Team organizational concepts are key to training data success. From the company's viewpoint, what is changing? Are the differences between training data and data science reflected in the organization? What new organizational structures are needed? Even if you are already in an AI-centric organization, there are training data–specific nuances that can help accelerate your progress.

New Expectations People Have for the Future of AI

Right now, the de facto standard process of getting a bunch of people together to annotate seems akin to an old typing pool, an army of people, doing relatively similar, predetermined work, in order to translate from one medium to another. A "typing pool" of annotation, outsourced or otherwise, can work for some contexts, but is clearly inefficient and limited. What happens when the strategic direction, local priorities, or individual work goals change? Where is the room for autonomy within business units, teams, and at the individual level?

Instead, what if people think of AI as part of their daily work? What if every new or updated process gave, first, a thought to AI transformation? What if every new application thought first about how the work could be captured as annotation? What if every line of business thought first about how annotation factored into their work? Explore before and after examples in Table 7-3 to consider the effects of the AI transformation.

Table 7-3. Before and after comparison of new AI expectations for staff

Before	After
Hire a new separate pool of workers (usually outsourced)	Your existing experts and data entry folks (primarily)
One-off projects, separate, one-time efforts	Part of daily work, like using email or word processing
"Bolt-on" mindset	"AI First" assumes AI will be present or demands it, integrated systems, aiding in the flow of existing work
AI being "pushed" onto people	People "pulling" AI into the org

Yes, people who are training AI instead of just doing their normal job may demand a higher wage. However, the return on capital is still a great deal better, even when paying a higher wage for one person, if that person, when combined with AI, is as productive as two or three people.

During the transition, it's natural that you still may need additional help. Depending on your business needs, there may be valid use cases that require outsourcing. As with any labor, there is a need for a spectrum of expertise in annotation. But the key difference here is that the annotation is seen as normal work, and not a separate project for "those people" to do.

The need for a range of talent requires a different approach when it comes to where you get that talent. A pool of workers exclusively hired "to annotate" without any other context at your business is distinctly different from any other level of worker hired into your business directly.

Another way to conceptualize this is to imagine a company of 250 people. Hiring 50 people overnight would be a massive deal. Yet that same company may think it's OK to hire 50 annotators. Try to see it more as truly hiring 50 people into your company.

You may already be thinking, there are a few areas that would be good targets, and/or "Well this sounds great, but I just can't see a way to annotate the such and such process." The reason I brought up this typing pool concept is that, when AI is a bolt-on, after-the-fact process, there will always be such hurdles. The more the organization is, instead, pulling AI in, seeing annotation as a new part of their day-to-day work, being directly involved in annotation, the more opportunities will come.

Sometimes Proposals and Corrections, Sometimes Replacement

A simple example of integrated proposal and correction that you may have perhaps already used is email. For example, in Gmail, it will prompt you with a suggested phrase as you're typing. That phrase can be accepted or rejected. Additionally, the suggestion can be marked as "bad" to help correct future recommendations, proposals, predictions, etc., as shown in Figure 7-5.

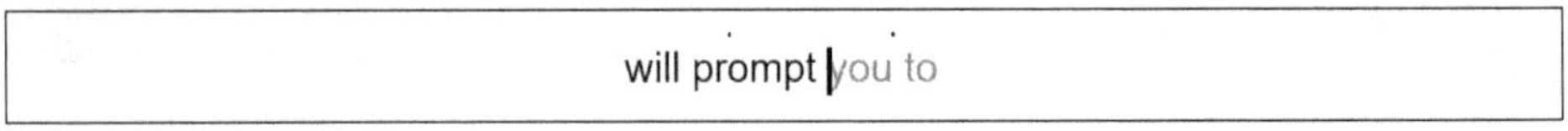

will prompt you to

Figure 7-5. Example of AI proposal to user

This highlights an important consideration for all of the products you buy as you go forward. It also circles back to the theme of using AI to make someone more productive, rather than directly replacing them.

Upstream Producers and Downstream Consumers

Training data work is upstream to data science. Failures in the training data flow down to data science (as shown in Figure 7-6). Therefore, it's important to get training data right.

Why am I making a distinction between training data and data science?

Because there are clear differences between the day-to-day responsibilities of people *producing* training data and data science people *consuming* it.

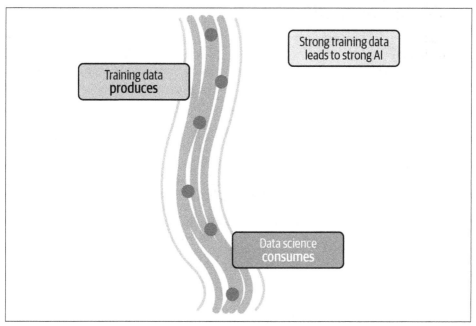

Figure 7-6. Relationship between production of training data and its downstream use by data science

Producer and consumer comparison

I think of the relationship between training data and data science as a producer and consumer relationship. See Table 7-4 for a comparison.

Table 7-4. Comparison of producing and consuming training data

Training data—producer	Data science—consumer
Captures business understanding and needs in a form usable by data science Converts unstructured data to structured data	Creates models that map fresh data back to business needs
Responsible for annotation workflow	Uses annotation output
Manages dataset creation, curation, maintenance	Uses datasets, mild curation
Supervises validity of outputs from data science to the business problem	Prediction outputs validity to training data
Example: KPIs (Key Performance Indicators): % of business need covered by data % of annotation rework required volume, V\variety, and velocity of annotation Depth of annotation	Example KPIs: Model performance, e.g., recall or accuracy Inference runtime efficiency GPU/hardware resources efficiency

Producer and consumer mindset

To a data scientist, the thoughts are often along the lines of "What datasets do we already have for x?" or "If I just had x dataset then we could do y." There is almost a short circuit, so to speak, where a project starts, and the moment there is an idea for something, the question is "how quickly can we get a dataset for this?"

An analogy is almost like, say I'm hungry and I want to eat something. I want to eat it now. I don't want to worry about the farm crop, or the harvester, or anything like that. There's nothing wrong with that—we all need to eat, but we must realize the distinction. That is, even if farms and farmers might not come to mind when I'm hungry, of course, they are relevant to my situation. In our case, the farmers are like the producers of the training data, and they are equally important to training data.

The more one learns about training data, and the more emphasis is placed on production, the further one gets from concerns about how to actually use the data.

As an illustration of this, I had a conversation with a leading training data production director who was trying to figure out how to get a specific type of rotated box. I suggested annotating as a 4-point polygon, and that the box can be provided based on the bounds of the polygon. This was a surprise to him—he had thought of box and polygon as two totally distinct forms of annotation.

The deeper you get into the data science world, the more the focus becomes on the literal consumption of the data, rather than the production of it. That data science consumption step is usually relatively clear-cut, e.g., a one-time mapping. It's the human interactions that produce the data, the ongoing challenges around the production of AI data at scale—that production is the important thing to focus on for a training data expert.

Why is new structure needed?

Bad data equals bad AI. AI that is dangerous, misfires in commercial settings, and becomes wasted, or even a negative investment.

Second, as part of the goal of AI transformation, there must be a principally responsible individual to lead the charge. While a VP or CEO can also play this role at the strategic level, the Director of AI is responsible to execute this strategy. This Director of AI role is naturally critical to the success of AI projects. The new role will help the Director build out the needed mindsets and tools around data. To do this effectively, they must have the appropriate power within the company.

Third, as the volume of people involved balloons, the simple reality is that this is a team of teams, and with people of many distinct characteristics. Even a small team (by large company standards) may likely have at least one or two production managers, and twenty to fifty annotation producers. In some cases, everyone using an AI tool is producing some form of human alignment. In a large organization, there may be hundreds or even thousands of annotation producers, all working to align multiple AIs and AI tools to commercial interests.[1]

It's an army of people to be managed. Strong leadership is required, and the budget to go along with it.

The budget

One of the most perplexing things about data science and training data organization is how to manage budgeting. Often, only a very small team of data science professionals are needed relative to a much larger volume of people doing data production. From a cost perspective, the data production cost may be an order of magnitude greater than data science. Yet somehow, the data science line item is often the top-level item.

If your overall objective is to maximize your AI results, an improved setup is:

AI/ML Structure:

- *Data production*: Everything to do with the production of data, including inserting and updating datastores, training data, human alignment, data engineering, and annotation
- *Data consumption*: Systems, tools, and people that consume data, querying datastores, including data science
- Supporting and other roles

This concept may seem straightforward, so why aren't teams already set up with this structure?

Historically, one reason for this has been in part the large hardware cost that data science is responsible for. While it's expected that this AI training and running cost will decrease with time, the cost needs to be allocated as the team is built and scales. Additionally, from a divide-and-conquer perspective of resources, the data science team is already burdened with managing the hardware cost, so it makes little sense to further cloud their mind with training data concerns. Rather, dedicating budget and team headcount to a robust data production and data consumption model will maximize this skill set on the team and ultimately deliver better AI.

---

1 As a quick note on outsourcing, even if a percentage of the annotators are outsourced, there is still such a substantial budget involved that it's probably better to think of them as part-time or even full-time equivalents.

The AI Director's background

There are a few skill sets that are essential to consider when looking for someone to fill the important role of AI Director:

- This is a people-leadership role.
- This is a change agent role.
- The person must be in tune with the business needs.
- Ideally, the person is able to traverse multiple departments of the company; perhaps is already a corporate-level analyst.
- There must be some level of technical comprehension to facilitate discussion with engineering.

What the background need not be:

- Formal education requirements. This is more a school-of-hard-knocks role. In practice, though, this person may hold an MBA, under or graduate scientific degree, etc. Most likely, they will also be up-to-date on the latest refresher and online courses in machine learning areas.
- A "Data Scientist." In fact, the more data science background the person has, the more risk that they'll focus on the algorithmic side over this new creative, human-focused side.

The Director's budget will need to be divided into two main components, people and tools.

Director of Training Data role

Given the aforementioned team structure, it is most ideal to have a *Director of Training Data* position established early.

This person can, for example, report to the VP of AI, the VP of Engineering, or the CTO. Even if this role is baked into some type of Director of AI role, the level of responsibility stands.

Ideally, this is a full-time, dedicated person in this Director role. In case it is not possible to have a dedicated person, then the role can be taken on by someone else.

Figure 7-7 illustrates both the Director of Training Data's responsibilities and sample descriptions of the key team member roles. These are not meant to be complete job descriptions, just highlighting some of the key structural elements of the role. The roles within the AI Center may be combinations of part-time or full-time positions. It's assumed that the engagement with the rest of the org will be mostly on an as-needed basis.

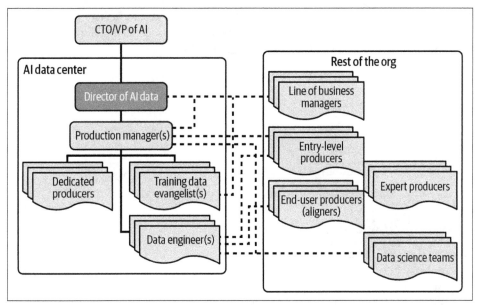

Figure 7-7. New org chart example

Note that Figure 7-7 refers to "teams" or plural "engineers." Of course, your org will not match this chart exactly. Think of it as a starting point. Each box can be thought of as a *role*. One person may play all or most of the roles.

AI-focused company modifications

- The Producers will still vary depending on the company. For example, the Experts may be the end users, or they may be part-time and external, but still named partners not a generic "pool."
- May have fewer or zero evangelists.
- Line of business manager may be the overall Product Manager.

Classic company modification

These may have fewer dedicated producers and more evangelists.

Spectrum of Training Data Team Engagement

Ideas around the primary organization pulling AI into its process is not mutually exclusive with the idea of there being a team or department for training data. In an AI-mature organization, the team may be primarily acting as advisors, staying on top of the latest trends, and maintaining the overall tools.

What's right will depend on your specific org; my main intent here is to convey the general spectrum, and that the idea of a separate team is needed, even if they aren't the ones doing the literal production of the annotation data.

The team will have three main areas of responsibility:

- Advisory and training
- Maintaining tool sets used by producers in other teams
  - Data ingest/access
  - Support for annotation production
- Actively managing production of annotation data

Dedicated Producers and Other Teams

Dedicated producers are direct reports of the production manager. This is for cases where the volume of work is such that the person's full-time job is annotation, and they are not attached to any specific business unit. Again, long-term, this may be rare, but it's a reality for teams getting started, transitioning to this structure, and for various projects where no other production capabilities are available.

For the sake of simplicity in the diagram, outsourced teams can be thought of as dedicated producers.

Organizing Producers from Other Teams

Producers in other business units run the spectrum, from entry-level to expert. End users, who may be different, may also produce their own annotations. Usually, end-user annotations are more "by accident" as part of using the application, or providing some form of minimal feedback.

Some of these titles speak for themselves, but others may be a bit more opaque. The following section walks through how each role contributes to better AI. Let's dive in!

Director of AI data responsibilities

Chiefly, this person is responsible for the *overall production and consumption of data*, with a focus more on production as the primary priority.

This includes the following main areas of responsibility:

- Turning the line of business needs into successfully produced training data.
- Generating work for training data production by mapping business needs to training data concepts.
- Managing a team of production managers who facilitate the day-to-day annotation production.
- Managing "Evangelists" who work with the line of business managers to identify training data and AI opportunities. This is especially relevant when it comes to feasibility concerns regarding annotation. Of the various ideas proposed by, say, a line manager, only a handful may actually be cost-effective at that moment in time to annotate.
- Managing the training data platform.

Besides general efficiency and visibility into annotation work, this person must map the productivity in annotation back to the business use case. *The Evangelist may do this too, with the Director being the second line.*

And there are some additional things the Director of AI should attend to:

- Managing the consumption of data via coordination with data science (DS) or also managing DS directly.
- Acting, indirectly, as a check and balance on data science (or internal DS teams), acting as a check on the business results and on the output of DS, going beyond purely quantitative statistics.
- Normal director-level responsibilities, including if applicable: headcounts, profit and loss responsibility, KPIs, supplier and vendor relationships, reporting, planning, hiring, firing, etc.

Naturally, the director can fill in for most any of the following roles as needed.

Training Data Evangelist

This role is that of an educator, trainer, and change agent. Their primary responsibilities include two areas:

- Working closely with the line of business managers to identify key training data and AI opportunities.
- Working "ahead" of production managers, establishing the upcoming work and acting as the glue between the line of business managers and the production managers.

In a company focused exclusively on AI products, their focus should be training:

- Educating people on the best usages of modern supervised learning practice.

In a classic business, there will be more breadth:

- Educating people in the organization on the effects of AI transformation. In practical terms, converting interest to actionable annotation projects.
- Recruiting annotators from that line of business. On a practical level, this would be about converting an employee doing regular work into someone who, say as 20% of their job, is capturing their work in an annotation system.
- Training. In the context of part-time annotators especially, this person is responsible to explain how to use tools and troubleshoot issues. This is distinct from the production managers, who are more geared toward training full-time annotators. This is because you train a doctor differently than you train an entry-level employee.

Training Data Production Manager(s)

This person is chiefly concerned with being a taskmaster for actually getting annotation work completed:

- Interfacing with data science to set up the schemas, setting up the tasks and workflow UIs, doing the admin management of training data tooling (generally non-technical).
- Training annotators.
- Managing the day-to-day annotation processes.
- In some cases, basic data loading and unloading can also be done by this person.
- When in change-management type discussions, this person is responsible to explain the reasonableness of annotation work to people new to the matter.
- Using data curation tooling.

Annotation Producer

Annotation Producers usually fall within two buckets:

- Part-time, which will increasingly become part of everyone's job to a degree
- Full-time, dedicated and trained people
  — May be newly hired people or reassignment from existing work

The primary responsibilities of Annotation Producer include the following:

- Implementing day-to-day alignment of AI to business needs
- Surfacing issues in the AI schema
- Daily front-line usage of AI tools, and tools that use AI

Data Engineer

Data Engineers have the following responsibilities:

- Getting the data loaded and unloaded, the technical aspects of training data tools, pipelines setup, pre-labeling, etc.
- Especially important is organizing getting data from various sources, including internal teams
- Planning and architecting setup for new data elements
- Organizing integrations to understand the technical nuances of capturing annotations

Data Engineers interface regularly with the data science team(s). Previously, this was out of necessity—often, projects would simply not work without close planning. Now, there are growing standards around AI datastores that make it possible to interface more on conceptual levels, rather than hashing out every implementation detail.

Historical Aside

There are a few reasons the Production concept wasn't needed before:

Classic machine learning the datasets already existed (even if messy), so there was no need to "Produce" a dataset.

Earlier efforts were more separated from day-to-day business goals. This meant there was more of a rationale to do one-time planning, one-off projects, isolated projects, etc. As AI transformation moves into the mainstream of your business, this separation becomes an artificial barrier.

Use Case Discovery

How do we identify viable use cases? What is required and what is optional? In this section. I provide a basic rubric to identify valid use cases. Then I expand into more context to help further identify good use cases.

This section is organized from the most concrete to the most abstract:

- Rubrics for good use cases
- Example use case compared against the rubric
- Conceptual effects, second-order effects, and ongoing impacts

The simplified rubric may be a go-to in day-to-day work, while the rest can act as supporting knowledge. While you are welcome to use the rubrics exactly as they are, I encourage you to think of everything here as thought starters, merely an introduction to thinking about use cases for training data.

Rubric for Good Use Cases

At the highest level view, a good use case must have a way to capture raw data, and at least *one* of the following:

- Repeated often
- Involves experts
- Adds a new capability

The more of these things that are present, likely, the greater the value of the use case. A rubric is shown in Table 7-5. Note that this rubric is focused on training data-specific use case questions and is not a generic software use case rubric.

Table 7-5. Rubric for good use cases

Question	Result (with example answers)	Requirement
Can we get the raw data?	Yes/No	Required
Is it repeated often?	Yes/Sometimes/No	At least one of these is required
Does it involve an expert?	Yes/Sometimes/No	At least one of these is required
Does it have a positive effect on multiple people's existing work?	Yes/Sometimes/No	At least one of these is required
Does it add a new capability?	Yes/Sometimes/No	At least one of these is required

That's it! This can be a go-to reference rubric, and to support it when needed, you can use the more detailed rubric in Table 7-6.

A key note before we dive in, some of these use cases may be a major project that has a large scope. Naturally, the level of scope will be relative to your specific case. If a use case will be a big effort relative to your organization's resources and capabilities, it is up to you to determine its applicability.[2]

Detailed rubric

Now that we have the general idea, let's expand on it. In Table 7-6, I provide more detailed questions, especially expanding on and differentiating the new capability concepts. I add examples, counter examples, and some text on why it matters to help.

Table 7-6. Detailed use cases rubric

Test	Example	Counter example	Why it matters
Is the data already captured? Or is there a clear opportunity to add more sensors to capture the data (in its entirety)? (Required)	Existing documents (e.g., invoices), existing sensors, adding sensors	Car dealer in-person sales interaction	• Getting raw data capture is a required step. • If you can't get the raw data then the rest doesn't matter!
Does it involve experts?	Docter, engineer, lawyer, some specialists	Grocery shopping	• Makes a constrained resource available to more people (and potentially more often, and in more situations). • Expert opinions are of high value. • More readily available data, often already in digital form.
Is the work repeated often? Many times per minute? Hourly? Daily? Weekly?	Automatic background removal/blur Customer service and sales Administrative document review		• Already will have a well understood pattern (at least by humans). • Likely already is relatively well constrained. • Often repeated tasks, viewed in aggregate, have a high value. • Existing raw data may be already being captured.

Adds a new capability use case

Sometimes the goal is to unlock new use cases, beyond augmentation or replacement. (Add a new capability.) A use case rubric to help make great use cases when the goal is to add a new AI capability is shown in Table 7-7.

---

2 I have also left off general use case/project management questions like "Scope acceptable relative to our resources?"

Table 7-7. Use case rubric when the goal is to add a new AI capability

Test	Example	Counter example
Is the work done rarely because of expense? Would it be of great value to increase the frequency?	Inspections	Stadium construction
Can we turn an approximate process into a more exact one? Does this process gloss over something because doing it in more depth is currently impractical? Are we currently substituting what we really want to figure out for something more generic? Would improving this process's accuracy lead to more benefits than harm?	Fruit ripeness, produce mold or bruising detection, dented can detection An airport detection system that just detects metal, versus something that can detect very specific threats	Loan underwriting
Is there something that gets completely skipped because it would take too long or is otherwise impractical (e.g., because of volume)? Would it be of great value if we could do it? Anything is relatively time-intensive (even if it happens rarely).	Analysis of video meetings and sales calls Porn detection in video uploads, comment moderation Insurance property review[a]	

[a] May take a while per house, but may only need to be done once per year or decade.

The main thing to distinguish this use case is that it's something that otherwise wouldn't happen. For example, a bridge may be inspected annually currently, but it would be impractical to inspect it daily. So an automatic bridge inspection system will add a new capability. This is a good use case, even though currently it is not repeated often (annually), and may or may or may not involve expert labor directly. For example, the actual inspectors may be looking for cracks and measuring them, while the engineering analysis is still done by someone else. Either way, it would be impractical for anyone to inspect it every day.

As we transition from no AI to AI-assisted work, or even relying solely on AI for some cases, there are naturally going to be overlaps, e.g., cost duplications. Also, the goals of adding AI vary widely, from new capabilities regardless of cost, to improving performance or economy.

AI has the potential to add a new capability that otherwise would have not existed. However, that new capability may be more expensive than even a similar human capability, and may not perform even at the level of a human initially. It also might be generally not economically viable.

Repeating use cases

a. Don't jump to assuming replacement—first think of augmentation.

b. Normal coding is fine for forms. Rather, think "what does a human do when they get the information?"

c. To get a good idea, look at how many "repeat" roles there are in the company. Are there thousands of people doing roughly the same thing? That's a great place to start.

Specialists and experts

All work involves some degree of specialization and training. Instead of offering "low-hanging fruit" in specific cases, here are some of the areas that often have the most opportunity (not necessarily the easiest). The expert case is generally meant to signify something that is otherwise of great difficulty to get. Of course, what *expert* means will be up to you; my own mental model is something along the lines of "a skill that takes 5–10 years, after normal education, to get to a basic level of proficiency, or an area that is so cutting-edge as to limit the pool of available people."

Evaluating a Use Case Against the Rubric

Here we present an example use case in some detail and then compare it to the rubric.

Automatic background removal

Have you recently done a video call and noticed someone's background was blurred? Or maybe you already use this feature yourself. Either way, most likely you have already interacted with this training data–powered product.

Specifically, when you take a videoconference (e.g., Zoom) call, you may be using the "background removal" feature (Figure 7-8). This turns a messy, distracting background into a custom background image or smoothly blurred background—seemingly magically.

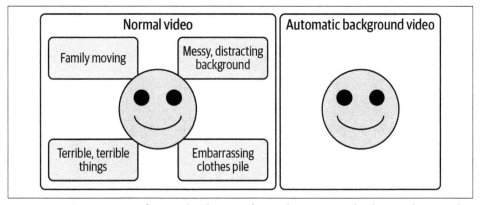

Figure 7-8. Comparison of normal video to video with automatic background removal

For context, this used to require a green screen, custom lighting, and more. For high-quality productions (like movies), often, a human has to manually tune the settings.

So, how is training data involved?

First, we must be able to detect what is "foreground" and what is "background." We can take examples of video, and label the foreground data as shown in Figure 7-9. We will use that to train a model to predict the spatial location that's "foreground." The rest will be assumed to be background—and blurred.

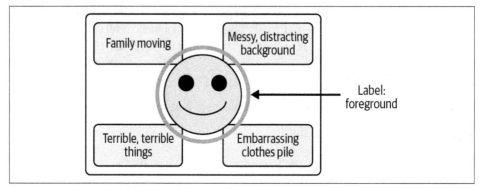

Figure 7-9. Example of labeling foreground

The point here is that the model is figuring out what patterns make up the "background." We aren't expressly having to declare what a messy pile of clothes looks like.

If we constrain the problem to assume that only humans will be in view, we could take an "off the shelf" model that detects humans and simply use that.

Why does a feature like this matter?

- Iterates more equality. Now it doesn't matter if I have a fancy background or not.
- It improves privacy and meeting efficiency. It helps reduce impact of disturbances (e.g., someone entering the edge of the calling area).

If this seems really simple—that's the point. That's the power of training data. Something that used to be *literally impossible* without a green screen becomes as easy as labeling videos.

Now a few gotchas to keep in mind:

- Getting a performant model to do "pixel segmentation" at time of writing is still somewhat challenging.

- There are public datasets of humans that already do a fairly good job. However, if you had to label a "Zoom video calls" dataset from scratch it would be a tremendous amount of work.

Evaluation example

Overall, the Automatic Background Removal use case scores fairly well in our evaluation, as shown in Table 7-8. It covers the requirement of being able to get the data. And gets a huge yes in the "repeated" category. Plus, depending on what subset of the use case you want to evaluate, it potentially avoids having to have an expert and adds a new capability. For example, I can't imagine being able to easily get a custom background in the middle of a cafe or airport without this.

Table 7-8. Example of using the evaluation rubric on the Automatic Background Removal use case

Question	Result	Requirement
Can we get the data?	Yes, the video stream is already digitally captured.	✓
Is it repeated often?	Yes, a single video call may remove thousands of background frames, a single user may have multiple calls per day, and there are many video callers.	✓
Does it involve an expert?	Sometimes. It takes some expertise to set up a green screen effect, but is far from expert medical or engineering knowledge.	–
Does it add a new capability?	Sometimes. It was possible before to get a green screen, but in cases like traveling, even a person who owns a green screen would be unable to use it, so in that case adding a new capability.	–

Use Cases

The conceptual effects area has a small list of use cases. For the sake of this rubric and timing, I dive only into a single use case deeply. My intent is to convey the conditions with which any use case would be good, and provide tools to think about the value add and overall effects. Given the breadth of potential use cases, I believe this is much more valuable than attempting to iterate a list of all known use cases; many such up-to-date lists can be found with online searches.

Conceptual Effects of Use Cases

This is a fairly similar idea to above, but from a slightly different perspective. In the preceding section, we were looking at this from the perspective of "What is a good use case?" Now I am looking at "What are these use cases doing?" I have also included some surface-level, obvious second-order effects—effects caused by adoption of the technology outside of the scope of the technology itself. I suggest using Table 7-9 as merely a starting point for thinking about the second-order effects of these concepts.

Table 7-9. Conceptual use case examples and second-order effects

Concept	Examples	Second-order effects
Relaxing constraints on the problem itself (on a prior solved problem)	• Green screen -> Any background. • I must be of a certain age to drive. • Spellcheck - > Grammar check.	• My background is not used to evaluate my candidacy for a job interview. • "Taking the kids to x" takes on a new meaning if the parent isn't driving. • Expectations around correct grammar change (in addition to correct spelling).
Replacing or augmenting routine work	• Human counting sheep -> automatic counting of sheep. • Human driving -> Car driving (parity with human performance). • Human routing communication to a department -> automatic communication reporting based on intent (sales, support, etc.).	• The meaning of work changes. • Millions of jobs will be created and shifted. People will be required to learn new skills. • The suburbs may extend further out. • A company that is not using AI effectively will have a worse cost structure than one that is (e.g., the same as not using digital effectively).
Making humans "superhuman"	• Airport security scanning. • Sports analytics. • Self-driving (accident reduction). • Acting as a "second set of eyes" on routine medical work.	• Airport security may become more effective and faster (Here's hoping!). • Intensity of sports may increase since expectations of elite levels of coaching are extended to more people. • Accidents may become more rare and even more newsworthy.
Making a constrained resource available to more people[a] (or without as many limits)	• Radiologist's time. Prior, a radiologist could only see as many people as they had time in the day; now an AI medical system can assist a radiologist in seeing a larger number of patients.[b] • This also removes the geographical limits. • Self-driving (more mobility because of lower cab fare due to shared resource).	• Medical care may become more accessible. • The meaning of a "second opinion" may change. • New dangers will appear, e.g., increased groupthink, data drift, decreased weight on human expert opinion.

[a] This is similar to the "Relaxing Constraints" but the examples are fairly different, so we'll keep them in separate categories.
[b] There are many limitations to the implementation details of this, of course.

Ongoing impact of use cases

At its core, the idea is that training data is an easier way to encode human knowledge into a machine.

The cost to "copy" human understanding approaches zero. While before, a radiologist's time was a scarce resource, it will become abundant. Before, a green screen was only in a film studio; now it's on my smartphone anywhere in the world.

I use the word "understanding" to differentiate between something that can be written down in a book, and something that up until now, could only exist in the human mind. Now machines, too, can form "understandings."[3]

This has the following follow-on implications:

- It is possible to dramatically increase the frequency of actions.

  For example, prior a visual bridge inspection may only be able to happen once a year or once a decade. Now, an analysis of a similar level could happen every few seconds.

- Processes that were previously "random" will become relatively "fixed."

  We all know car accidents happen. But eventually, they will be rare. What was previously a random process (we've all asked someone dear to us to "call me when you get home") will turn into an all but sure thing. (When there is one, it'll be international news! "The first car accident in the last 24 months just happened!")

- Previously impossible things will become possible.

  For example, putting a "dentist in your pocket"; eventually, you will be able to point your phone's sensors at your mouth and get a level of insight that previously would have required a dental visit.

- Increased personalization and effectiveness of "personal" assistance will increase.

The New "Crowd Sourcing": Your Own Experts

Who is annotating and training data culture go hand in hand. The more people who understand training data, the more opportunities will rise up. The more your own staff and experts are involved, usually, the higher the quality will be. There are a few other things to keep in mind:

- Buying into the "training data first," a data-centric mindset, is one thing.
- Who is actually doing the annotation is a different decision.
- Reframing your existing teams as "The Crowd" will help you get higher-quality data at a greater quantity.

By leaning into your own staff, you can get better results while controlling costs.

---

3 Technically, this is the "higher dimensional space," which we are unable to reasonably represent visually. Many machine learning models have hundreds of dimensions, while we can only reasonably graph four (space (x,y,z) and time (t)).

Key Levers on Training Data ROI

These factors will ultimately determine the "shelf life" of training data and its ability to be leveraged into tangible productivity improvements:

Talent, or who is annotating
> To quote someone who managed a large team of annotators: "The biggest determinant of annotation quality is the person who created the annotation."

Degree of training data culture
> This can be a step function–type delta. Either people are aware something can be turned into training data or they aren't.

When it comes to AI annotation talent, if you have to pick one, lean toward quality over quantity.

What the Annotated Data Represents

Thinking about your annotation in that context further supports the need to have a dedicated business unit focused on it. It also showcases the need for awareness at all stages of the buying process. For example, if you are buying a system to automate something, and the vendor is responsible for the annotated date, what does that mean for your future? This is what's at stake:

- Your business know-how, trade secrets, processes, and competencies
- The massive labor investment you needed to make to create, update, protect, and maintain it
- Your key to staying competitive during the AI transformation

Trade-Offs of Controlling Your Own Training Data

Arming your team with the knowledge they need to tackle training data in house can bring many benefits to your business:

- Better quality by leaning into existing experts and domain knowledge
- A better cost model by repurposing or reframing existing work
- Creating a durable, shareable, and reusable library of training data
- Control over both the economics and quality of the output
- More internal management overhead, potentially less flexible
- Requires organizational knowledge and change efforts

The Need for Hardware

First, let's get some sticker shock out of the way. Big, mature AI companies spend tens to hundreds of millions of dollars on AI computing resources (e.g., GPUs, ingestion, storage). This means that hardware cost is a key consideration.

Second, training data is your new gold. It's one of your most important assets. People may be your most important asset, but this is a literal embodiment of people. It's important. How do you protect it?

Realistically, if you aren't controlling the hardware, there is very little you can do to protect that data. What happens if there is a contract dispute? What if the vendor's controls aren't as good as you thought? You have your key business data and records under your control—and training data must be the same way. To be clear here, I don't just mean to say the annotation tooling. Any data that any of your vendors are capturing that's used for training AI must be considered here.

This means that while a SaaS solution may be OK for getting started, proof of concept, etc., for ongoing training data requirements, the hardware costs, and degree of importance to the company, are too great to not take control of it.

Practically speaking, if you leave any type of prediction, annotation automation, etc., in the hands of a vendor's server (unless it's being run on the client) means you will be actually negotiating a huge amount of hardware costs hidden inside the annotation tooling cost.

Common Project Mistakes

Under-resourcing is especially prevalent here. I have seen a number of solo professionals, from pathology to dentistry, be curious about AI. While this curiosity is naturally great, the reality is that you need a strong team effort and substantial resources to build something that will be even a true prototype level, or in production.

There are a few common roadblocks to getting started:

- Project is under-resourced. A single doctor is unlikely to make a general-purpose AI, even for their niche area.

- Mistakes on volume of data needed. For example, a major dental office's x-rays for all time for all patients may be substantial, but on its own, still probably is not enough data for a general-purpose dentist AI.

- Most AI projects have a very, very long time horizon. It takes on the order of months to years to build reasonable systems. And often, the expected lifetime and maintenance is measured in years or even longer.

Modern Training Data Tools

The effective usage of training data tools can make orders-of-magnitude differences. The first step is to gain awareness of what high-level concepts exist. If you have read all the way through, you have already taken a big step in this direction; getting this book and other material in front of your team is a great way to help accelerate this process.

Software can help provide guardrails and encouragement to a transformation, but it is only a part of the overall transformation.

Training data–focused software is designed to scale to the breadth and depth of complexity that's needed to actually capture your business processes with the fidelity required. There's a big difference between tagging a whole image, versus a polygon with complex attributes, that has gone through a dedicated review process.

Training data software has come a long way, and has had millions of dollars of investment since its humble origins. Modern training data software is more in line with an office suite, with multiple complex applications interacting together. While perhaps not quite as complex line for line, directionally, it will be in a few years' time.

Think Learning Curve, Not Perfection

There is a tendency to seek out perfection and familiarity in training data software, especially if an existing early team happens to be familiar with a certain pattern. The simple truth is that all software has bugs. The other day, I used Google search, and it duplicated the menus and search results. That's a product with oodles of engineering effort over decades! If we think back to early computer applications, they had terribly obscure UIs. People had to learn many concepts to perform simple tasks. The same general principle applies here.

The other truth to understand is that these application and use cases are continuing to increase in complexity. When I first started this, I could provide a demo of most of the key functions in half an hour. Now, even if I scope the demo to a specific persona (like an annotator), and a specific media type (like an image), it still may take half an hour! A complete end-to-end coverage would be days, similar to how even a basic training course, for someone who's never used a word processor or spreadsheet, would take days.

So how do we tackle this learning curve?

Having deeper and broader discussions
> While UI design and customization is important, over-focusing on these elements can cause us to miss key points. If the SMEs are so busy that they can't take some basic training (or just time) to figure out how to use the application, then realistically, are they going to provide quality annotations?

Customization and configuration

Further, while off-the-shelf software should absolutely be the starting point, we have to recognize that there will always be the need for some degree of customization and configuration.

Training and new knowledge

New training and knowledge are required, from leadership to frontline staff, and from concepts to specific tooling.

New Training and Knowledge Are Required

Let's discuss some specifics on what training is needed—people training in this case, not training data.

Everyone

- The introduction: high-level overview of what Supervised AI is and how it relates to your specific business.
- Clarity of AI's role. For example, if it's the case that AI will generate Proposals, and Humans will do alignment.
- The effect of AI on work. For example, that AI supervision leads to more productive and interesting work.
- The task: To bring up ideas of processes that are ripe for this type of supervision.

Annotators

- The basics of annotation tooling. To continue the office analogy, knowing how to use annotation tooling will be the modern equivalent of learning word processing.
- More in-depth training, such as, in part, reading this book, and further training on sensitive issues like bias.

Managers

All of the above, and:

- Knowing what questions to ask regarding new and updated processes.
- Understanding how to identify financially viable training data opportunities.
- Reflecting on productivity goals in this new world of AI annotation. Every moment of work done that's not captured in annotation is a moment lost.

Executives

Will need to reflect on the company organizational structure and consider elements such as the creation of a new training data unit

Should germinate, nurture, and guard the culture around training

Will need to carefully consider vendor choices in relation to securing future AI goals

How Companies Produce and Consume Data

Who is making the software that consumes training data? Who is producing the data?

Here are the major themes I have seen in practice:

- A software-focused company produces an AI-powered product, and the bulk or all of the training data is also produced by this company. The firm releases the software to consumers, or another firm buys the software and is the end user.

- A company with in-house training data production capabilities creates the software for its own internal use. Often, this may involve leaning on external partners or very substantial investment.

- A software firm produces an AI-powered product, but leaves the bulk of the training data to be produced by the end user buying the software.

The only case where the end-user company is not involved in the training data production is 1. In general, that either shifts the core competence of the business to that software provider, or it means there is a relatively static product being provided. An analogy here would be buying a "website" that can't be updated. Since few people want a website they can't update, in general, the trend will likely be toward companies always being able to produce their own training data in some way.

From the executive viewpoint, in a sense, the most key questions are: "Do you want to produce your own data?" and "Do you want AI data to be a core competency of your business?"

Trap to Avoid: Premature Optimization in Training Data

Optimizing your training data too early in the process can bring challenges. Let's explore this further.

Table 7-10. Premature optimization, warning signs and how to avoid it

Trap	How it happens	Warning signs	Avoiding it
Thinking a trained model = Done	1. Effort is taken to train a model. 2. It sort of works and people get excited. 3. Assume it just needs some small adjustments. 4. Realize it's far from done.	• The "Trained Model" is discussed as the end goal. • Ongoing continuous annotation is not discussed. • Iteration is talked about, but in the context of a limited time window.	• Educate people that the goal is to set up a continual improvement system, not a single one-off model. • Discuss up front what performance level is good enough to ship version 1. For example, for self-driving, some have taken the approach of "equal to human is good enough."[a]
Committing to a schema too early	1. Use up a lot of resources annotating. 2. Realize the labels, attributes, overall schema, etc. do not match their needs. E.g., were using bounding boxes and realize they need keypoints.	• Schema does not change significantly between early pilot work and more major work. • Schema was determined with minimal data science involvement. • "Final" schema was determined without proof that achieving success will solve downstream problems.	• Expect the schema to change. • Try many different schema approaches with actual models to see what actually works—don't assume anyone has the right background to know the answer in advance. • Ask: if the model made perfect predictions, does this actually solve our downstream use case? E.g., if it perfectly predicts this box on rebar, will that solve the overall problem?
Committing to automations too early	1. Look at how many resources human annotation may take. 2. Look for automation solutions. 3. At first are happy with automation results. 4. Realize the automation isn't quite doing what they that it was doing.	• Unrealistic expectations arise, e.g., automation expected to virtually solve annotation. • Automation explored without involving data science. • Automation plans being discussed in detail before any significant human annotation work has been done. • A reduction in management mindshare in training data under false assumption that automations handle it.	• Realize it doesn't take much human annotation to start getting a directional understanding of needs. • Use minimal or zero automations until you have done enough manual human annotations to well understand the domain. • Do more human annotations than you think you need. After, you will be in the best position to choose the most effective automations. • Think of automations as an expected part of the process, not a silver bullet.
Wrongly calculating volume of work	1. Look at overall dataset size. 2. Project how many annotations. 3. Assume will need all of them.	• Assuming all available data needs to be annotated (data can be filtered for most valuable items). • Not considering ongoing accumulation of data or production data. • Not considering diminishing returns, e.g., that every further annotated item adds incrementally less value than the prior item.	• Get a large enough sample of actual work to understand how long each sample usually takes. • Realize that it's always going to be a moving target. For example, the work per sample may get harder as the models get better.

Trap	How it happens	Warning signs	Avoiding it
Not spending enough time with tooling Not getting the right tools	See "Tools Overview" on page 37.	• Overfocus on "getting a data set annotated" instead of what the actual result is going to be. • Unrealistic expectations; treating it more like a fly-by shopping website than a serious new suite of productivity tools.	• Realize these new platforms are like "Photoshop walks into a bar and meets a gruff Taskmaster who moonlights as a Data Engineer." It's complex and new. • The more powerful the tool, the more the need to understand it. Treat it more like learning a new subject area, a new art.

a I'm not commenting if this approach is right or wrong.

No Silver Bullets

There are no silver bullets to annotation. Annotation is work. For that work to have value, it must be literally labored at. All of the means to improve the productivity of annotation must have a grounding in that real-life reality.

Training data must be relevant to your business use case. To do this, it needs the insights from your employees. Everything else is noise or in-depth expert, situationally specific concepts.

Instead of looking at some of these optimizations as true gains, rather see missing out on them as being functionally illiterate.

This caution is important because, being such a new area, the norms are poorly established, making a clear understanding of the history, future goals, and latest conceptual frameworks that much more important.

Culture of Training Data

Getting everyone involved with training data is at the heart of AI transformation. In the same way that the IT team doesn't magically re-create every business process on their own, each line of business manager brings a growing awareness of the capabilities of digital tools and what questions to ask.

One of the biggest misconceptions about modern training data is that it is exclusively the realm of data science. This misconception really holds teams back. Data science is hard work. However, the context in which that hard work must be employed is often confused. If the AI projects are left as "that's the data science team's responsibility," then how likely do you think success company-wide will be?

It's very clear that:

- Non–data science expertise, in the form of subject matter knowledge, is the core of training data.
- The ratio of people who are SMEs to data scientists is of vast fields to a single grain.
- Data science work is increasingly becoming automatic and integrated into applications

The great thing here is that most data scientists would be happy if other people knew more about training data. As a data scientist, I don't want to worry about the training data for the most part, I have enough of my own concerns.

Not everything, at least right away, is a good candidate for training data. Being able to recognize what will likely work and what won't work is a key part of this culture. The more bottom-up the ideas for this, the less likely it will turn out to be something that appears easy, but is actually terribly hard, and fails.

New Engineering Principles

The first step for getting a better data engineering setup in place is to recognize the intersection of the business need, and the expert annotators, with the day-to-day concerns about data. That will lead to needing an AI datastore that can combine all of your BLOB, schemas, and predictions in one place:

Separating UI/UX concerns of annotation experiences
For example, creating UI/UXs embedded into existing and new applications.

Making the training datastore a central component
To run a web server, we use standard web server tech. It's the same with training data. As the complexity and investment in this space continues to grow, it makes more and more sense to shift that training data into a dedicated system, or set of systems.

Abstracting the training data from data science
A simple break point here is dataset creation. Instead of a dataset being something that's static, if we allow training data to be created on its own accord, and data science to query it, we get a much more powerful separation of responsibility. Of course, in practice there will be some interplay and communication, but directionally that provides a much cleaner scope of responsibilities.

Summary

The age of AI is here. Are you ready? Human guidance is the foundation of AI transformation. Data-centric AI is about creating new schemas, capturing new raw data, and doing new annotations to build AI.

To recap, some other key points we covered in this chapter:

- Appoint a leader like a Director of Training Data to drive your AI strategy, oversee data production, and interface with data consumers.
- Identify high-impact AI use cases using rubrics focused on frequency, experts, and new capabilities.
- Rethink your annotation talent strategy to leverage your existing experts as annotators.
- Implement modern training data tools for annotation, workflow, validation, and infrastructure.

Next I'll cover automations, and we'll look at some case studies.

CHAPTER 8

Automation

Introduction

Automation can help create robust processes, reduce tedious workloads, and improve quality. The first topic I'll cover in this chapter is pre-labeling—the idea of running a model before annotation. I'll cover the basics and then step through more advanced concepts like pre-labeling just a portion of the data.

Next, interactive automations are when a user adds information in order to help the algorithm. The end goal of interactive automations is to make annotation work a more natural extension of human thought. For example, drawing a box to automatically get a tighter location marked by a polygon feels intuitive to us.

Quality assurance (QA) is one of the common uses of training data tools. I cover exciting new methods like using the model to debug the ground truth. Other tools automatically check base cases and look at the data for general reasonableness.

Pre-labeling, interactive automations, and QA tools will get you far. After covering the foundations, I'll walk through key aspects of data exploration and discovery. What if you could query the data and only label the most relevant parts? This area includes concepts like filtering an unknown dataset down to manageable size and more.

I will touch on data augmentation, common ways it's used, and cautions to be aware of. When we augment data, we derive new data based on the existing base information. From that viewpoint, it's easier to think of the base information as the core training data and then the deriving process—augmentation—as a machine learning optimization. So while a portion of the responsibility here is outside the scope of training data, we must be aware of it. Simulation and synthetic data have situation-specific uses, but we must be up front about the performance limitations.

We have a lot to unpack and experiment with in this chapter. Let's get started by taking a closer look at the project planning process and techniques that are commonly used today.

Getting Started

High labor costs, lack of available people, repetitive work, or cases where it's nearly impossible to get enough raw data: these are some of the motivations to use automations. Some automations are more practical than others. I first outline common methods used, followed by what kind of results you can expect and not expect. This is rounded out with the two most common areas of confusion about automations—fully automatic labeling and proprietary methods.

I wrap this section by looking at the costs and risks. This section is a view of how the concepts map together, and ultimately of how they actually help your work. It can also help direct your reading and act as a reference to quickly look up common solution paths.

Motivation: When to Use These Methods?

When working with training data, you'll likely encounter problems that automation can help with. Table 8-1 covers some of the most common problems with automation-focused solutions.

Table 8-1. Reference of problem and corresponding automation solution

Problems	Solutions
• Too much routine work • Annotations too costly • Subject matter expert labor cost too high	• Pre-label
• Data annotated has a low value add	• Pre-label • Data discovery
• Spatial annotation is tedious • Annotator day-to-day work is tedious	• Interactive automations • Pre-label
• Annotation quality is poor	• Quality assurance tools • Pre-label
• Obvious repetition in annotation work	• Pre-label • Data discovery
• It is near impossible to get enough of the original raw data	• Simulation and synthetic data
• Raw data volume clearly exceeds any reasonable ability to manually look at it	• Data discovery

Check What Part of the Schema a Method Is Designed to Work On

Automations work on different things. Some work on labels, some attributes, others spatial locations (e.g., box, token position), others more general concepts outside of the schema. For example, object tracking is generally oriented toward spatial information and doesn't usually help as much with the meaning (labels and attributes). When possible, I will highlight this, otherwise research the specifics in depth, as it usually comes down to the implementation details. As an example of this impact, a method that offers a 2x improvement on the spatial location will be of little importance if 90% of the human time is in the attributes, defining the meaning of what is actually present. This is relevant to all automations that impact annotations directly.

What Do People Actually Use?

With so many new concepts and options, it can be easy to feel overwhelmed.

While these methods are always changing, here, I highlight the top methods I have seen people use successfully and that are the most broadly applicable across all projects. The two questions I will most try to answer:

- What are the best practices?
- What do you need to know to be effective and successful with training data automations?

Commonly used techniques

These approaches may be commonly used; however, this does not mean they are easy, always applicable, etc.

You must still have sufficient expertise to use the approaches effectively—expertise you will gain in this chapter. These are the most frequently used techniques:

- Pre-labeling[1]
- Interactive automations
- Quality assurance tools
- Data discovery
- Augmentation of existing real data

---

1 For experienced folks, grading production predictions is also rolled into this concept.

Domain-specific

These techniques depend on the data and sensor configuration. They can also be more costly, or require more assumptions about the data:

- Special purpose automations, like geometry and multi-sensor approaches
- Media-specific techniques, like video tracking and interpolation
- Simulation and synthetic data

A note on ordering

In theory, there is some order to these methods; however, in practice these methods have very little universal order of usage. You can pre-label as part of a data-discovery step, during annotation, or much later to grade production predictions. Sometimes, data discovery is only really relevant after some significant percent of the data has been labeled. Therefore, I have put the more common methods toward the start of the chapter.

What Kind of Results Can I Expect?

Being a new area, it's common for there to be few expectations about what these methods can do. Sometimes, people invent or guess at the expected results. Here, I unpack some of the expected outcomes of using these methods.

Table 8-2 covers these methods. First, it shows the method itself. Next is the expected result when well implemented. And finally, the "Does not" covers the most common issues and confusions.

Table 8-2. Overview of automation methods and expectations

Method	Does:	Does not:
Pre-labeling	• Reduces the less meaningful work. • Usually within a single sample, e.g., an image, text, or video file. • Shifts the focus to correcting odd cases. • A building block for other methods.	• Solves labeling entirely. • A human must still review the data and do further work with it.
Interactive automations	• Reduce tedious UI work. • For example, drawing a box, and getting a polygon drawn tight around an object in the box.	• Completely eliminate all UI work.
Quality assurance tools	• Reduce manual QA time on ground truth data. • Discover novel insights on models.	• Replace human reviewers entirely.

Method	Does:	Does not:
Data discovery	• Keeps human time focused on the most meaningful data. • Avoids unnecessary similar annotations.	• Works well in cases where there is already a well-functioning model and the main goal is grading it.
Augmentation	• Provides a small lift in model performance.	• Works without having base data.
Simulation and synthetic data	• Cover otherwise impossible cases. • Provide a small lift in model performance.	• Work well for cases where raw data is already abundant or situations are relatively common.
Special purpose automations, like geometry and multi-sensor approaches	• Reduces routine spatial (drawing shapes) work by using geometry-based projections. • Some limited methods can work largely independent of humans; others usually require labeling of the first sensor, and then the rest is filled in.	
Data-type-specific, like object tracking (video), dictionary (text), auto border (image)	• Reduce routine work by utilizing known aspects of the media type.	

Common Confusions

Before we dive into this, let's address two of the most common areas of confusion, automatic labeling for novel model creation and proprietary automatic methods.

"Fully" automatic labeling for novel model creation

There's a common myth that I hear repeated often: "We don't need labeling, we made it automatic," or "We automatically label to create our model (no humans needed)."

There are many techniques closely related to this, such as distilling larger models into smaller ones, or labeling specific portions of data automatically, etc. Many related techniques are explained in this chapter. These techniques are a far cry from "we completely automated data labeling for new model creation."

Consider that when an original AI model makes a prediction (generates a label) that is different from creating that "original" AI model in the first place. This means that some form of non-automatic human labeling must be involved to create that original model. Further, GPT techniques, while again related, don't yet solve this either for original model creation. A GPT model can "label" data, and you can fine-tune or distill a large model to a smaller one, but the alignment of the "original" GPT model still requires human supervision.

While this may eventually change, it is out of the scope of commercial concerns today. If we get to a place where we have true fully automatic labeling of arbitrary data for novel model creation, then we will have achieved the elusive idea of artificial general intelligence (AGI).

Proprietary automatic methods

"Use our method for 10× better results"—sometimes, this is a myth, and sometimes, it's true.

The intent of this chapter is to convey the general methods and concepts that are available. As you will see, as these methods stack with each other, it is possible to get relatively dramatically better results than by purposefully going as manually and slowly as possible.

However, there are a vast number of concrete implementations. The single most common theme I have seen with secret vendor-specific approaches is that usually they are very narrow in scope—for example, it may work for only one type of media, one distribution of data, one spatial type, etc. It can be difficult to verify in advance if your use case happens to meet those assumptions. Statistically, it's unlikely.

The best way around this is to be aware of the general approaches as explained in this chapter, and trend toward tools that make it easy to run the most up-to-date research and use your own approaches.

User Interface Optimizations

All tools have a variety of user interface–based optimizations. They are things along the lines of copy and paste and hotkeys. They are worthwhile to learn. As these are specific to each tool, and more of a "trivia" thing, I do not cover them in the book.

Outside the scope of "standard" UI expectations, there are interactive optimizations, which are specific named concepts and covered separately following.

Risks

All automations, even those operating as expected, introduce risk. Here are a few that are specific to annotations:

Lack of net lift
> This is when automation doesn't actually help. It is surprisingly common, especially with automations that involve user interface.

Worse results
> Automations can make results worse. For example, the super-pixel approach can lead to blotchy annotations that are less accurate than a traced polygon. Don't assume the failure state is equal to manual annotations.

Cost overruns

As detailed in the next section, automations have many costs, including time to implement, hardware costs, human training cost, etc.

Method-specific risks

Every method is unique and comes with its own unique risks.

Even seemingly similar methods can have different effects; for example, correcting a pre-labeled box is different from a pre-labeled polygon, and different again from an attribute classification.

Treating it too casually

Sometimes, these automations feel like just a little time-saver, or some kind of obvious thing. We have to remember that every automation is like a mini system. That system can have errors and cause problems.

Trade-Offs

Automations can sometimes feel like magic. Wouldn't it be nice if the system could just do x for us? This magic feeling shrouds some of these methods with an almost religious zeal. While many automations can and do improve training data results, they need careful trade-off analysis, planning, training, and risk analysis to be used safely and effectively.

It's important to note that most of these techniques have shifted over time, so they are now less about being incremental savings opportunities, but rather have become degrees of standard expectation. This is similar to how a company that uses a word processor doesn't think of it as savings in comparison to a typewriter anymore.

It's quite incredible how quickly a dataset can be improved by a careful combination of automation methods. However, automations are not magic, and all have trade-offs. In this section, I'll provide some general rules to help evaluate some of the conceptual trade-offs of automations:

- The conceptual nature of automations
- Setup costs
- How to benchmark well
- How to scope the automation relative to the problem
- How to account for correction time
- How methods stack, which can also stack costs and risks

It's worth keeping in mind that no matter what automation is used (GPT included), subject matter expertise is still required.

Nature of Automations

The general guidelines for using automations is that the humans should always be adding true value with every annotation. If the annotation is too repetitive, it's a strong signal that a change is needed, and automation is an option among other options, like changing the schema.

Some automations leverage knowledge from another model, something akin to knowledge distillation. Now, we cannot hope to get a better new model by directly running an existing model in a loop, but we can leverage the existing knowledge to make the work of the annotator more focused on adding true net value. This is part of why automation can't "annotate" for the purposes of new model training, but can be a powerful assistant to a human.

It is difficult to scale a project without the use of some forms of automation. Avoiding automations entirely may make projects intractable financially, in completion time, or both. A project without automations may be hard to get beyond a basic threshold of getting started. An analogy here is having CI/CD processes. It's possible to work without them in theory, but realistically, you need them to scale your team and projects in a commercial setting.

Setup Costs

All automations require some form of setup. Setup can take a few different forms:

- Training on how to use them
- Implementation efforts
- Time needed to understand the technical risks

Even the ones with the most minimal of setup require a degree of training and understanding around their assumptions. While this may seem obvious, it's often glossed over, and must be remembered.

How to Benchmark Well

Previously, it was common to compare automations against manually completing 100% of the project.

Nowadays, completing an entire project with 100% manual annotation is rare. By manual annotation, I mean no model review, no pre-labeling, no UI assistance features, none of the methods in this chapter. Yet, automation methods are often presented as a percent savings relative to this mythical 100% manual annotation.

Instead of thinking of it as a comparison to manual annotation, think of it as comparing to best practices—and best practice is to use these automation methods appropriately.

There are a few benchmark questions that you can use to assess your organization's use of automation:

- Are we reasonably aware of all commonly available methods? Of the ones relevant to our case, what percent of them are we using? For example, if there are four methods available, and you are using all four, then at the strategic level, you are already there.

- By survey, find out how unique do annotators feel the work they are doing is? Do they feel they are making small adjustments or adding foundational value with every annotation? This can still be quantitative, but the point is to get away from lower-level metrics that are easily gamed or noisy.

- What is the ROI of our automation methods? For costs, consider all supplies (e.g., hardware), vendor costs, administration, and data science costs. For the benefit, if possible, try the actual next best alternative. This requires being clear-eyed about the comparison.

How to Scope the Automation Relative to the Problem

Imagine you have a lawn to mow. It's a small front lawn about the size of a parking spot. You can get a push mower or a ride-on mower. The ride-on, being about as big as the parking spot, is so effective it can pretty much just start up and instantly cover the entire lawn.

Naturally, the ride-on mower is orders of magnitude more expensive than the push mower, and it requires storage, supplies, gas, maintenance, etc. So while once in place, it will mow the lawn the fastest, its startup costs and ongoing costs far outweigh the benefits.

Setting a good scope of automations is important. A few questions to consider:

- Does the automation require technical integration? Can it be done from a UI or through a wizard?

- What are the expected startup and maintenance costs?

- Is an off-the-shelf method adaptable to our needs? Do we need to write a specialized automation?

If this sounds like a project management discussion, that's good, because that's what each automation is—a project. Obviously, a giant lawn mower is too large for a small lawn. So what are the right size metrics for training data?

Correction Time

While it might not always be factored in, time spent performing corrections should be accounted for. An automation that will cause a lot of need for corrections may not really save time in the end. If a spatial speed-up method is being used, try just drawing the polygon points directly by tracing the object. When you account for correction time, determine which method was actually faster.

I have seen research papers that reference "clicks." They say something like: We reduced the number of clicks from 12 to 1; therefore, it's 12 times better. But in reality, there is another tool that allows you to trace the outline, meaning that there are no clicks, and it's just as fast to trace the outline as it is to fiddle with the errors the automation makes.

As you can see in Table 8-3, the magic tool still takes some time to draw. In this case, it takes more time to correct, meaning that while it appears to be faster, it's in fact slower.

Table 8-3. Example comparison of time to draw versus time to correct

Tool	Time to draw	Time to correct
Trace outline	17 seconds	N/A
Magic tool automation	3 seconds	23 seconds

Subject Matter Experts

It's also worth understanding that virtually all of these methods still end up requiring a similar level of subject matter expertise. An analogy is that collectively, the tooling represents something akin to a word processor. It saves you having to buy ink, to mail letters, etc. But it doesn't write the document for you. It automatically formats the characters you instruct it to write.

To be clear here, requiring a subject matter expert does not mean the SME has to literally do every annotation themselves. They may instruct others on how to annotate specific sections, use tools, etc. Further, there may be cases where it's possible to contrast degrees of expertise.

Using SMEs does not mean you have to jump to the most expensive or most expert person in the world. As a concrete example, there may be many medical radiology annotations that can be done by a technologist, assistant, resident, etc. Referring to a radiology SME does not presume that it must be a full radiologist, or radiology professor.

Consider How the Automations Stack

Many of these improvement methods stack together. This means that you can use multiple methods together to achieve a positive effect, as detailed in later sections. The risk here is that, similar to taking multiple medications, the interactions and side effects are poorly studied. This can create hard-to-determine errors, make troubleshooting more difficult, and incur multiple costs.

Pre-Labeling

Pre-labeling is inserting predictions into training datasets. Pre-labeling is popular, and one of the most universally usable automation concepts.

Three of the most common goals of pre-labeling are to:

- Reduce the effort to improve the model (standard pre-labeling)
- Correct or align existing model results (QA pre-labeling)
- Add custom annotations on top of existing baseline predictions (custom data pre-label)

It is usually best to clearly define what is a pre-label and what is a human annotation, and to understand well the intent behind pre-labeling. One of the biggest risks with pre-labeling is a "feedback loop" where successive iterations increasingly generate more and more uncaught errors. Pre-labeling is usually focused on labels within a sample, for example, bounding boxes or segmentation masks in an image. Pre-labels can also be used for data discovery; see "Data Discovery: What to Label" on page 260 for more.

Standard Pre-Labeling

Let's imagine we have a model that is sort of good at predicting faces. It usually gets good results, but sometimes fails. To improve that model, we may wish to add more training data. The thing is—we already know that our model is good for most faces. We don't really want to have to keep redrawing the "easy" examples. How can we solve this?

Pre-labeling to the rescue! Let's unpack this in Figure 8-1. First, some process runs, such as a model prediction. Humans then interact with it. We then typically complete the "loop" by updating our model with the new data.

Initially the user reviewing the data is shown all five predictions. A user may directly declare a sample to be "valid," or the system may assume that any samples not edited are valid by default. However it's marked, the net result is the second stage—four correct examples and one incorrect one that needs to be edited.

The annotator must still verify the correct samples. The verification time should be less than the time it would take to create it in the first place in order for this approach to be effective.

The general theme of this method is that over time, it focuses the human user on the hard cases, the "net lift."

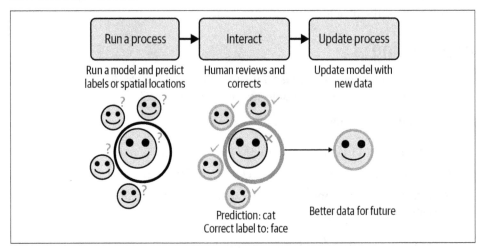

Figure 8-1. Pre-label overview

Benefits

Pre-labeling is popular because it works on nearly anything, is conceptually clear, and is relatively easy to implement. There are some clear advantages to pre-labeling:

Universal works on anything that can be predicted
Standard pre-labeling is not tied to any specific method. It works across nearly any domain. Essentially, anything that a model, system, or existing process is generating can be used as a pre-label. This means it works on nearly any schema concept, from What something is to Where it is.

Conceptually clear and easy to implement
Pre-labeling is relatively easy to implement and understand. An existing system (usually a model) generates a prediction, and that's shown to the user to review, add more data, etc. Because these systems can generate the predictions at any time, and then be reviewed at a later time, it is relatively straightforward to store, load, and show this data.

Quality assurance is baked in (sometimes)

In some cases, such as when a model is generating a prediction, pre-labeling acts as a quality assurance control. Specifically, it ensures that people are looking at the actual predictions, as opposed to some other automation method where the entire process is more hidden. Making this data visible creates a relationship to the actual model that can be very valuable.

What's the difference between model quality assurance and pre-labeling?

There is a similar technical process to load the data, and the actual correction mechanisms are usually fairly similar in the UI. With quality assurance, generally the intent is to still improve future models, and this is again similar to pre-labeling.

Caveats

Pre-labeling still requires a system to generate the initial data, can create unexpected errors, can be misleading in terms of efficiency, and can potentially become confusing in more complex cases. There are a few more cautions to keep in mind:

Requires an existing relevant model (or system)

The primary caveat is that you need an "initial" model or system to generate the data. In general, pre-labeling is a "secondary" step in the process. This can be offset in that some methods and contexts require very small amounts of data— for example, maybe even a handful of images and you can start—but it's still a blocker. Transfer learning does not solve this, but it does help reduce how many samples are needed to start using this new model.

Can introduce new error modes, bad feedback loops, and diminishing returns

Instead of looking at data with "fresh eyes," it may be easy to miss something that "looks good," but actually is wrong. It can also reinforce errors through bad feedback loops. Pre-labeling can have diminishing returns when the schema difficulty is not well aligned.

Sometimes can be slower or misleading

In the context of literal annotation, editing may be slower than creating a new thing for some cases. If a model is predicting a complex spatial location and it requires lots of time to correct, it may be faster to simply draw it from scratch. Zooming out to a more conceptual level, pre-existing models can be misleading in terms of automation value. For example, an existing model predicts a fishing boat, because it was already trained on fishing boats. This may seem obvious, but it's best to be clear what data the model was already trained on to understand what is actual new work versus existing data. In the context of standard pre-labeling, it's best to only be correcting the models we plan to use, otherwise what's correct and not can become confused quickly. However, as I'll speak to next, you can also separate your data value add from the existing model.

Pre-Labeling a Portion of the Data Only

You can pre-label a portion of the data with a different model. The portion of data labeled could be things like a top-level label, or just the spatial location, etc.

Usually a model used for pre-labeling a portion of the data is not appropriate to be used as the actual primary model. This is because either it's too narrow (e.g., just spatial location, like segment anything) or because it's too large, compute intensive, or unpredictable, like a massive GPT model.

An extension of pre-labeling is to use a model to predict something it knows, and use it as a starting point for predicting something unknown. An example of this is dividing responsibilities (e.g., spatial information and attribute updates). For example, using an existing model that we don't plan to update for spatial predictions, and focusing our responsibility and effort on adding new information, e.g., attributes. In some approaches, a "micro model" gets discarded. The micro model is only used within the scope of annotation automation. This approach can help by saving time on the dimension it's good at, while the humans add meaning or other needed dimensions for the "real" model.

For example, imagine we have a face detector that's great at predicting faces but it doesn't know anything else. We can run that model first, get the spatial locations of the faces, and then add the new information, like happy, sad, etc., as shown in Figure 8-2. While it may seem subtle, it's a big difference from standard pre-labeling. We aren't specifically trying to correct the spatial locations; we are strictly using it as a time-saver to get to labels and attributes.

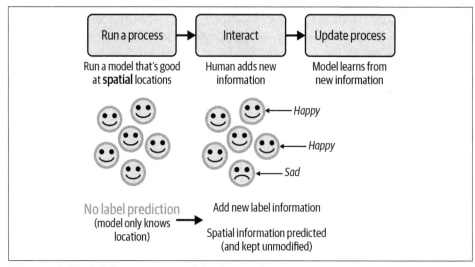

Figure 8-2. Pre-labeling a portion of the data only

Imagine a sports broadcast of a basketball team—identifying and tracking every player in each frame manually would be a lot of annotation work. There are fairly good "people detectors" that can run, as well as tracking algorithms, etc. These are responsible for determining the spatial location. Then, the human annotation can be focused on the relevant items, like what action the players are doing—things that a generic tracking/detector algorithm doesn't provide. Note that these approaches are also known as: micro model, custom data on existing model pre-label, and sometimes knowledge distillation.

Use off-the-shelf models

Because our goal here is to add new information (not to correct an existing model), we can use off-the-shelf models. This means that even if we don't have any training data for our Happy/Sad model, we can get started with an existing face detector. This can mean faster setup time. Also, because we aren't directly concerned about improving the model, if built-in models are available, they can be used directly without drawbacks.

Clear separation of concerns

Pre-labeling the specific portion of the data either helps speed up the process in the moment—or it doesn't. This is a clear separation of concerns relative to the more general pre-labeling, where the separation of concerns of in-the-moment annotation and overall data changes is more blurred.

An off-the-shelf model may not exist, or it may require upgrading to be useful. Sometimes they are not directly helpful for improving spatial detections or are not very interactive, and fairly static.

Note that some GPT-related pre-labeling processes may be less clear-cut or may have different trade-offs. This is a developing area.

The "one step early" trick

Speaking to the idea of automation more generally, it's possible to get a prediction "one step early." This might mean a generic object or event detection without knowledge of the class.

Typically, this is combined with some other method, e.g., pre-labeling.

How to get started pre-labeling

The data ingestion chapter covers some of this in more detail, but here I review the steps required to get started:

1. Identify your existing models.

2. Map the data from the model to the training data system.

3. Create tasks using the existing data.

4. Annotate.

5. Query and use the data to retrain new models.

Next let's talk about a more dynamic concept, interactive automation.

Interactive Annotation Automation

Interactive automations are used to reduce UI work. Conceptually, this is where it's clear to the human what the correct answer is, and we are trying to get that concept into the system faster. For example, a user knows visually where the edges of an object are, but instead of tracing an entire outline, the user guides an interactive segmentation system. Popular examples include SAM (Segment Anything Model), box to polygon (e.g., GrabCut, DxtER), and tracking algorithms.

These work best when a user interacts with them, usually by adding information, such as a region of interest, in order to help an interactive algorithm generate something useful. In theory, this makes UI work a more natural extension of human thought. Or put another way, as users provide "interactions" (inputs), then we can derive "interactive" speed-up approaches. Back to the semantic segmentation example, what if I could just point the computer toward what general area of the image I'm looking at, and it would figure out the specific shape?

Note that interactive automation is a developing area, and the examples shown are meant to be illustrative, not definitive.

A major assumption here is that the user is standing by, ready to add that initial information and then review the result of the interactive process. This process is shown in Figure 8-3, and is essentially composed of three steps:

1. User generates initial interaction.

2. Process runs (usually on the order of seconds of time).

3. User reviews results.

The key test to help compare interactive automation to other methods is that it should be impossible for the process to run without some kind of initial user input. In contrast, full image or full video methods can be run as a background computation without direct user input. This is the crucial difference. Another difference is that usually it's not the model of interest actually running.

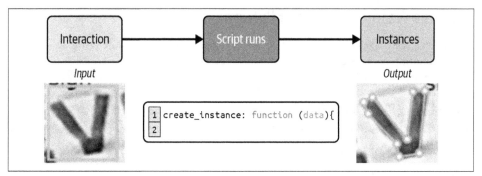

Figure 8-3. Conceptual overview of interactive automations

There are a few pros and caveats to consider when looking into interactive automation:

- Pros
  - For complex spatial locations, it can be extremely effective.
  - It usually does not require a trained model, or an existing model.
- Caveats
  - Can sometimes be "recreating the wheel." E.g., if you have a model that's good at detecting people, it's probably better to use that to start than to manually identify each person (even if it's a single click).
  - Often requires more user training, patience, etc. Sometimes, these methods can take so long to run that they don't really save much time.
  - To really get these methods working well, it often requires some form of engineering work over and above wiring input/output.

Creating Your Own

Some tools provide integrated methods, or off-the-shelf approaches. For those tools, you may not have access to modify how they work. In those cases, the following examples will serve as a guide to the inner workings behind the scenes. Diffgram provides a compiler to enable you to write your own automations. This means you can use your own models, the latest open source models, adjust parameters to your desire, etc.

Technical Setup Notes

For the scope of these examples, I'm using open source Diffgram's automation library.

The example code is all in JavaScript. For the sake of brevity, most of the examples are pseudocode. Please see the linked examples for full executable code examples.

What Is a Watcher? (Observer Pattern)

Any time a user does something, such as creating an annotation, deleting an annotation, changing labels, etc., we think of it as an *event*.

We do so at a semantically meaningful level for training data.

For example, a `create_instance` event is more meaningful than a regular `mouse_click` if I want to do something—such as running a model—after the user creates the instance.

How to Use a Watcher

We must first define a function that will do something upon a `create_instance` event, shown below. After we enable this script, and the user draws an example annotation, we see the example output in Figure 8-4:

```
create_instance: function (data){ // your function goes here }

create_instance: function (data){
    console.log(data[0])
}
```

Figure 8-4. Left, example of code editor for interactive annotation; right, logging annotation instance created

Interactive Capturing of a Region of Interest

Here is an example of code that captures the canvas region of interest based on a user's annotation. While this may seem fairly simple, it's the first step toward running an algorithm based on just the cropped area, as shown in Figure 8-5.

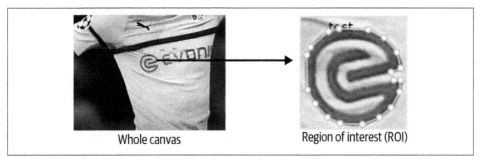

Whole canvas   Region of interest (ROI)

Figure 8-5. Example of whole canvas versus region of interest

The code to capture this looks like this:

```
create_instance: function (data){
  let ghost_canvas = diffgram.get_new_canvas()
  let instance = {...data[0]}
  let roi_canvas = diffgram.get_roi_canvas_from_instance(
    instance,
    ghost_canvas)
}
```

More generally, the idea here is to use some input from the user to preprocess what we feed to our model.

Interactive Drawing Box to Polygon Using GrabCut

Now, this example will take our region of interest canvas, run the standard OpenCV GrabCut algorithm and output polygon points, which will then be converted into a human-editable annotation.

Crucially, you can replace GrabCut() with a preferred algorithm of your choice. Here is the simplified pseudocode (JavaScript):

```
let src = cv.imread(roi_canvas);
cv.grabCut(src,//args)  // replace with your model or preferred algorithm
cv.findContours(//args) // because it's a dense mask
cv.approxPolyDP(//args)  // reduce to useful volume of points
points_list= map_points_back_to_global_reference()
// because roi is local, but we need the location relative to whole image
diffgram.create_polygon(points_list)
```

If you'd like to view the full example, you can find it here (*https://oreil.ly/dRa6W*).

Full Image Model Prediction Example

Zooming out from the directly interactive example, we can also run file-level models. You can imagine having a series of these models available, with the user perhaps choosing at a high level which one to run.

The high-level idea is to get the information from the user interaction and the raw media, and then run the automation.

An example in JavaScript could look something like this:

```
this.bodypix_model = await bodyPix.load()
let canvas = diffgram.get_new_canvas()
let metadata = diffgram.get_metadata()
segmentation = await this.bodypix_model.segmentPerson(canvas, {});
Points_list = get_points_from_segementation()
diffgram.create_polygon(points_list)
```

The full example is available here (*https://oreil.ly/U4gvf*).

Example: Person Detection for Different Attribute

Now, going back to a more simple example, we can also use these automations to simply run models.

Here, we used the BodyPix example in a Diffgram userscript. It ran on the whole image and segmented all the people. Here, we now select the person to add our own attribute, "On Phone?" as Shown in Figure 8-6. This is an example of using a model that's not directly our training data goal to avoid having to "recreate the wheel" for parts that are well understood. You can imagine swapping this model for other popular ones, or using your own models for the first pass here.

Figure 8-6. Example of spatial prediction automation with attribute value add

Quality Assurance Automation

There are novel training data quality assurance (QA) tools to debug ground truth data and reduce quality assurance costs.

Using the Model to Debug the Humans

The big idea here is that models can actually surface ground truth errors. Sounds crazy? So, the way to think about this is essentially that if a model has many examples of truth, it can overcome small errors. You could compare it to the intuition that sometimes allows us to spot errors on a quiz answer sheet because even though it says it's the answer, we think about it and feel confident that it's not the answer.

In practice, one way to do this with images is to use Intersection over Union (IoU). If the model prediction doesn't have a nearby ground truth, then either the model is wrong or the ground truth is. This can be surfaced by applications as a ranked list. Or put another way, this is about identifying the biggest delta between predictions and existing ground truth.

Automated Checklist Example

While this may sound simple, a checklist is an essential part of the QA workflow.

This can literally be a checklist for human review, or if possible, it can be programmed. Some tools may include some of these by default, but often the best checks are set up more as "test cases," where you define parameters relevant to your specific data (like a checksum). Here are some example parameters:

- Is the count of samples reasonable?
- Are the spatial coordinates reasonable?
- Are all attributes selected or approved by a human?
- Are all instances approved by a human?
- Are there any duplicate instances?
- (Video specific) Are there tracks where an instance is missing (e.g., in frame n, missing in frame n+1, reappears in frame n+2, and not marked as end sequence)?
- Are there issues with dependency-type rules (e.g., if class A is present, then class B should also be present, but class B is missing)? This is different from heuristics in that it's usually more about reviewing existing instances than generating new ones.

Domain-Specific Reasonableness Checks

Depending on your domain, you may be able to use or build specific checks. For example, if the class is a "person," and 95% of the pixels in the spatial location of the sample are all a similar color, that's unlikely to be correct. Or, you can check to see if the majority of pixels are in a different histogram distribution than what's expected (e.g., an object that is known to never have red pixels).

Data Discovery: What to Label

Data discovery tools can help us understand existing datasets, identify the most valuable data to add value to, and help us avoid working on data that is showing diminishing returns. This goes back to the concept of wanting every annotation to provide net lift, to make a positive incremental improvement in the model performance. In this context, "rare" or "different" data is simply any data that will help get you net lift.

Conceptually you can broadly think of these methods in three main categories:

- Human exploration
- Metadata based
- Raw data based

Human Exploration

All methods involve some form of human review at some point. Common human data discovery steps include the following activities:

- Visually exploring data in a UI catalog
- Querying data to identify subsets of interest (for example, "Carrots > 10")
- Reviewing the results of tooling (e.g., similarity search)

Human exploration happens at multiple stages in the process. This can be at the stages of raw data without annotations, data that has some annotations and may need more value added, subsets of data, and results from various tooling practices.

Human review is critical to catch mistakes in automation. Automations may make faulty assumptions about the data. For example, if you have a limited amount of data, reducing it in any way may cause unwanted results. While automation approaches for data discovery can be a very important part of the process, human understanding is always the first, second, and third step.

Raw Data Exploration

Raw data exploration methods are all about looking at the actual raw data.

An example is similarity search. One use case is to find new samples of a specific known sample type. This could be in the context of knowing you are low on annotated data similar to a given sample. The process is to run a similarity search on that sample to find more samples that are similar (but not yet annotated). Then, you create tasks to review those samples.

Another example of raw data exploration is identifying groups of data. For example, imagine you had both daytime and nighttime pictures. You could use a raw data exploration tool to look at that data and segment it into two batches. The system may not actually know that it's day and night pictures, but it will know by looking at the raw data there are two (visually) distinct groups. You can then sample from each, for example, 100 daytime samples and 100 nighttime samples. This is to avoid getting 1000 samples of a similar distribution as that shown in Figure 8-7. A good time to use this method is when your data contains many similarities and exceeds any reasonable ability to manually look at it.

Another use is identifying outliers. For example, you might flag rare samples that may not have enough volume to be understood correctly.

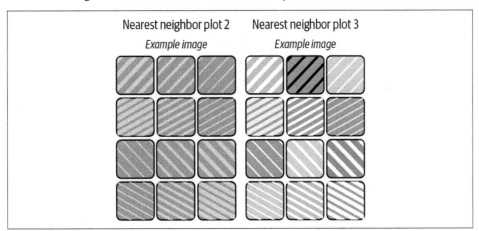

Figure 8-7. Example of similarity comparison

Metadata Exploration

Independent of the literal raw bytes of the samples, there is often attached metadata about the sample. For example, if you want to do something like this automotive company does in Figure 8-8, you can query based on metadata like the time of day, microscope resolution, position, etc.

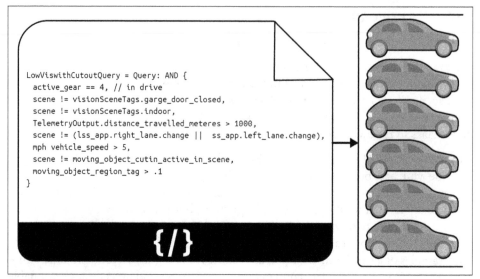

```
LowViswithCutoutQuery = Query: AND {
  active_gear == 4, // in drive
  scene != visionSceneTags.garge_door_closed,
  scene != visionSceneTags.indoor,
  TelemetryOutput.distance_travelled_meteres > 1000,
  scene != (lss_app.right_lane.change || ss_app.left_lane.change),
  mph vehicle_speed > 5,
  scene != moving_object_cutin_active_in_scene,
  moving_object_region_tag > .1
}
```

{/}

Figure 8-8. Example of query (pseudocode)

Make sure to add the metadata during insertion so it can be queried well. The example in Figure 8-8 is pseudocode. Diffgram has a good resource (*https://oreil.ly/ kCp-u*) if you'd like to explore query syntax reference further.

Adding Pre-Labeling-Based Metadata

An option is to use an existing network to pre-label at a course level, for example at the whole file level. For example, for a robotics company, say the intent is to label only "daytime" images, but for some reason the metadata of the data is unavailable. You can run a network to tag images as "nighttime" or "daytime," then only label the "daytime" images. In practice, there can be some rough edges to implement this, but in theory this can dramatically save time.

Augmentation

Augmentation is modification of real data. For example, skewing existing images, changing the brightness, introducing artifacts, etc. Augmentation is an active area of research, and opinions vary. Sometimes augmentation is proposed as a way to reduce training data effort. For example, popular libraries and processes often do a heavy amount of data augmentation, but what actual lift this is providing is sometimes overlooked. Generative models can also be used to augment data, and this is a rapidly developing topic at the time of writing.

To put it simply, augmentation is often a temporary crutch, and/or provides a relatively small lift, or even a decrease in performance. Think in terms of –10 to 10%. It can still be useful, and it's good to be aware of when and how it can be used.

In general, there seems to be a consensus toward doing augmentation at runtime, and more minimally than during the exuberance of the early approaches. The implication of this is that for most cases, it's not something for you to overthink, the same way we don't think about compiler optimizations for regular code.

Augmentation has its place, but we must remember that it's, at best, a small performance boost, not a replacement for real data.

Better Models Are Better than Better Augmentation

One general theme is that as the models and algorithms get better, augmentation becomes less effective. This may not always be true, but is a good rule of thumb.

A simple way to think of this is that there are nearly infinite subtle combinations of data values, be it text, pixels, etc. So, trying to train an ever-growing amount of subtly different data just to have a better understanding of what it is is a losing proposition.

Think about it from the human perspective—I don't need to see 100 different variations of essentially the same thing in order to recognize it.

To Augment or Not to Augment

The first thing to appreciate with any form of augmentation is that it significantly alters the playing field.

Next, the best evaluation of any augmentation method is what "net lift" it provides. This is sometimes trickier to measure than is first apparent.

For example, imagine a system that pre-labels instances. The spatial location is pre-labeled reliably. Expert users then add additional information, such as what type of crop disease it is. The "net lift" in this case is primarily time savings. So if the cost (literal cost, complexity, etc.) to set up the pipeline, run the model, etc., is low enough, then every spatial location that's pre-labeled is a direct time savings. In practice, if a robust model is available for this, or is already being run in production, then this is valuable.

Alternatively, let us consider some negative scenarios.

An approach (e.g., superpixels and similar) is used to more "quickly" label spatial locations. Except, it introduces artifacts and hard-to-correct errors. It's easy for it to look "OK" at a glance, so these errors go unnoticed until significant time is spent on the dataset. This may cause the need to rework and/or hard-to-understand production errors.

A pre-label approach that requires constant correction can also be problematic. Correction is ~3x slower than just drawing the spatial location correctly the first time. So, in the abstract, the pre-label must be very accurate to break even, and more so to provide net lift.

Any scenario in which an "external" model is used can cause difficulties. It can introduce bias into the new training data, polluting it, so to speak. As a practical example, imagine a person detector that appears to reliably detect people. It's used as the "spatial location" first pass, to which your customer labels are then added. If that detector has a bias, perhaps favoring an ethnic majority, it may be very difficult to spot, since it's "hidden." You may need an extra QA step expressly to check for this potential.

At a high level, the general rules of thumb for these augment approaches are:

- Explore with caution—there are always trade-offs, and remember that augmenting methods hide the trade-offs well.
- Consider the "net lift," taking into account costs such as the added complexity, reduced flexibility, literal compute/storage costs.
- Many of the approaches are very specific—they may work very well for one case, of one spatial type, etc., while failing on many others.

I have a few other general observations to keep in mind:

- The most effective options generally seem to favor spatial types over class labels.
- Generally, it's best to create a "seed" set, before exploring any type of "model"-based approaches there.
- Sometimes "augmented" training data is somewhat hard to avoid, e.g., in the case of "correcting" a production pipeline.

Training/runtime augmentation

As long as the augmentation method and process are reproducible, you may not need to store it alongside regular training data. Augmented data can be very "noisy" and is not typically that useful for human consumption. However, even when using readily available tools, implementing augmentation at training time comes with challenges, for example, memory and computational costs. In this context, moving it to runtime isn't always straightforward.

Patch and inject method (crop and inject)

The concept here is to take patches—meaning crops or subsections—of the data and create novel combinations. For images, you can imagine placing rare classes in a scene. This is sort of like a hybrid simulation where real data is simulated in different places.

This is still an area of research. In general, a model should be robust to these types of scenarios.

Simulation and Synthetic Data

Closely related to augmented data is simulation and synthetic data. Simulation and synthetic data techniques depend heavily on domain-specific contexts. Often, they are a blend of other arts and disciplines outside of training data.

All synthetic data methods still require some form of real data and human review to be useful. This means synthetic data shows promise, but is not a replacement for real data. That's why it's best thought of as the cherry on top of real data.

Let's get the big question out of the way real quick—does simulated data work? The super-short answer is "Yes, but probably not as well as you may be hoping."

The next question is, "Will it eventually work really well?" This one is actually easier—it's unlikely. And there's a simple thought experiment to reflect on this. To simulate the data, to the level of realism that is required, is on the artificial general intelligence level of hard problems. It's relatively easy to get photorealistic looking renders—but there's a big difference between those simulated renders and real life.[2]

Typically, simulated data, if used well, can provide a small lift in performance. It can also be sometimes used for situations where it would be impossible, or very difficult, to get the data beforehand.

Another way to think about this is that using a simulation to automatically create training data is somewhat akin to using a heuristic. Instead of humans supervising and saying what right is, we are back to trying to build a bottom-up type model—just one level removed.

Simulations Still Need Human Review

Shown in Figure 8-9 is an example from a major company that they presented publicly. They claimed it was "perfect" data. As you can see, there is a patch over the crosswalk—part of what they touted as being more realistic—but then they forgot to account for that in the training data generation, generating perfect crosswalk lines over the imperfect pavement.

---

2 Anecdotally, I have run this by both research- and industry-oriented people, and there's usually common agreement that it's true.

Were they trying to teach it that the black splotch was actually a crosswalk? This doesn't make sense. Of course, perhaps they were meaning to post-process it later to *project* a perfect crosswalk. And perhaps, some super-specialized network could theoretically do this. But at least from the training data perspective, this is just a plain error.

Figure 8-9 shows just a small example, but it gives the idea that assuming that a simulation will render perfect data is far from correct.

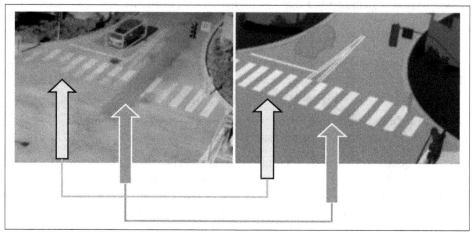

Figure 8-9. Example of training data that was claimed to be perfect but has clear labeling mistakes

In Figure 8-9, the left image shows the simulated image with faded white lines, and some white lines obscured entirely (by what appears to be pavement). The right image shows a training data rendering that has annotations that guess at where the line should be, ignoring the obscured and missing lines. If this data was actually trained on directly, it would be misleading to the model, since we would be saying that patch of pavement equaled a complete line, when clearly no line is visible.

For the impossible and rare cases, I would really think of it less as "automatically creating training data" and more as "automatically creating scenes, for which humans can then create training data."

For example, a system may be able to automatically identify which pixels are from which object. But, if the simulation is sufficiently random, that may not be relevant for events. In other words, it's important to think about what the simulation actually *knows*.

A way to think about this is if I have a simulation of shelves at a supermarket, those shelves will look similar, unless the simulation is engineered specifically to render different shelves. But what dimension will they be different in? Will those differences be relevant to the production data? Who is programming that in? Remember that even subtle shifts in the data can have huge effects on the result.

There are some clearly beneficial uses of simulation:

- Product images. If you already know what the product looks like, that should be a clear enough case for a shopping-type robot.
- Rare scenes of autonomous driving.

And there are some overall pros:

- Simulations can improve performance.
- Simulations can create otherwise impossible or rare cases.

And cons:

- The improvement is usually only by a small amount; think on the order of a relative 0–10%. There may be case-specific exceptions to this, so it's worth some research.
- Often, the fidelity of simulations is a lot further from the real world than may first appear. For example, what looks visually good in a video, may be drastically different pixel for pixel.
- Simulations are often costly to set up, maintain, and operate.

There are a few things to think critically about when considering simulations:

- What are we actually simulating? Are we simulating lighting conditions? Camera angles? Whole scenes?

Media Specific

Many automation methods work well for all media types. Here, I cover some of the highlights of how popular methods intersect with specific media types. Then I cover some common domain-specific methods. This does not cover all media types. The main intent of this section is to give you awareness of the relationship between specific media and automation types and provide an introduction to some of the domain-specific types.

What Methods Work with Which Media?

As you can see in Table 8-4, generally, most of these methods can stack well together. Of course, each method takes a degree of work and understanding. There is no requirement to use all of them. You can have a very successful project that completely skips a lot of these.

A ✔ indicates that usually the method will match. All methods require some degree of setup and have exceptions and cases where they may not work. For example, using pre-labeling with data discovery approaches for images is better defined than for video. I have included more extensive footnotes here to expand the rationale where required and keep the chart easy to read.

Table 8-4. Media formats commonly used with each technique

Methods/data type	Video	Image	3D	Text	Audio
User interface best practices	✔	✔	✔	✔	✔
Pre-labeling	✔	✔	✔	✔	✔
Interactive automations	✔[a]	✔	✔	Sometimes[b]	Research area
Quality assurance automations	Research area	✔	✔	✔	✔
Data discovery	Sometimes[c]	✔	Research area		
Augmentation[d]	Sometimes[e]	✔	Research area	Research area	✔
Simulation and synthetic	✔	✔	✔[f]	Research area	Research area
Media specific[g]	Varies—check method				
Domain specific Special purpose automations, like geometry and multi-sensor approaches	Few commonly used methods	Sometimes	Few commonly used methods	Few commonly used methods	Few commonly used methods

[a] May take more effort to set up.
[b] There are a variety of "real-time" model training concepts, but often this is more single-user focused.
[c] If the method requires conversion to images, this may prevent the use of video-specific automations. If possible, think in terms of discovering segments, and keep the video form, as it may improve compatibility.
[d] Augmentation methods tend to be domain specific. Opinions vary.
[e] Many augmentation methods are focused on images, so may not be applicable for event detection or other types of motion prediction tasks.
[f] Because most simulations are 3D by default.
[g] Like object tracking (video), dictionary (text), auto border (image). Because the methods are unique to each media type, double-check when choosing a specific method.

Considerations

Just because you are using a certain data type, e.g., video, doesn't mean that an automation will be available for that case. For example, maybe you have no problem identifying people in videos, but identifying specific interactions may be harder.

All tools have a variety of UI concepts that can help improve performance, including common things like hotkeys, copy and paste, etc. These methods sometimes get confused with other things, so I think of this as a separate category.

Media-Specific Research

An important thing to keep in mind is that some methods are more or less established for certain media types. For example, something that may be considered "standard" for images may be barely researched for text, or not even have a direct analog. For some people, this is not obvious because they may focus on one media type only (e.g., NLP or computer vision):

Video specific

If you are interested in *event detection*–type actions, some of these automations are not recommended. For example, object tracking would confuse a situation where you want to know exactly in which frame an event occurs:

- Object Tracking: The big idea here is to look at the data itself, and track an object throughout multiple frames. There are a variety of tracking concepts, many need only a single frame to provide an effective track forward through time.

- Interpolation: The idea here is humans create keyframes and the in between data is filled in, as shown in Figure 8-10.

Keyframe - 1   Interpolated frames 2 through 9   Keyframe - 10

Figure 8-10. Example of interpolation

Polygon- and segmentation-specific—auto bordering

I covered how to use auto bordering in Chapter 3. Here, I briefly want to state that using techniques like that to snap to edges, as simple as it may seem, is a great use of practical UI automations.

Text (NLP)-specific

Pre-labeling with GPT, heuristics (covered earlier in this chapter), and dictionary look ups are three of the most common methods.

Domain Specific

Many domains have specific approaches that work well for that domain only. For example, for real-world robotics, there is 3D geometry that often matters, whereas that type of geometry usually isn't relevant for all digital PDFs. There are many domain-specific options that can be implemented well. For instance, if you have multiple sensors, you can get a "two for one" deal by labeling one sensor, and using known geometry to estimate into the other sensors. Incorporating interpolation and object tracking can speed up video. Using dictionaries and heuristics can help with text. I'll walk through all of these at a high level so you are confidently aware of them.

Geometry-Based Labeling

In some cases, you can use the known geometry of a scene, sensors, etc., to use geometry-based transformations to automatically create some labels. This case is highly dependent on the specific context of your data.

Multi-sensor labeling automation—spatial

The main idea here is to use mathematical projections to assume where something is in space based off of where the sensors physically were. For example, if you have six cameras, it can be possible to construct a virtual 3D scene. Then, you can label one camera, and project that into the other five. Often this means projecting into 3D and back into 2D. Note that this does *not* require a 3D-oriented sensor like LiDAR or radar. But in general, it's considered easier if one of those sensors is present.

In a theoretically optimal case, you get something like a 5:1 return. E.g., for every one camera you label, you get labels in five other views. Keep in mind that you still need to review, and often correct, the projections, so in reality the return is more like 3:1 at best. Keep in mind it requires having multiple sensors, extra metadata, and the ability to project into 3D.

Spatial labeling

This is similar to the multi-sensor labeling, but is more focused on cases with a large extent of geometry-focused concepts, like lane lines.

Heuristics-Based Labeling

In general, heuristic labeling has been NLP-focused, and concentrated in two main areas:

- Dictionary-based labeling
- User-defined heuristics

This is a controversial area. If you have written a perfect set of heuristics, then why even train a machine learning model at all? The main problem is that the more well-defined a heuristic is, the more it looks like we are coding and doing work there, and basically reinventing feature engineering. Since one of the whole points of deep learning–based approaches is automatic feature engineering, this essentially is self-defeating. Now, this is not to say heuristics-based methods are without merit— far from it. There is some incredibly interesting research in this area, especially for text-based applications.

Summary

The theme, time and time again, is "Humans." From high-level data discovery tools, to framing specific metadata queries—human review takes center stage. When thinking about how to construct automated quality assurance flows and reasonableness checks, it's all about human knowledge when doing interactive automations to extend human thought. When humans are viewing and adding value to existing predictions with pre-labeling. When you, as the humans, are evaluating the trade-offs and risks of automations. There's always something to automate, and always more human steps needed to set it up, control it, monitor it, and maintain it.

You have learned an incredible amount of practical knowledge in order to understand the high-level risks of automations, the rewards seen of common methods, and many specifics, from media types to domains. You have had both a relatively comprehensive overview and the details to conduct further research for your specific challenges. Next, I'll dive into real-world case studies to help tie up all the concepts you have learned so far.

Overall, there are many high-quality approaches that will substantially automate your training data process. There are also limitations and risks to consider when implementing automation. I provided a high-level overview of the most common methods you can actually use today, what results to expect, the trade-offs involved, and how they work together. For the most popular methods, you now have a deeper dive into some of the specifics.

Next, let's dive into real-world case studies.

Case Studies and Stories

Introduction

Now I will share some exciting case studies and stories. I have aimed to provide a breadth of views, from engineering to annotation quality assurance. Stories from big companies to multiple sizes of startups, and education-centric data science competitions.

You may be just starting your AI path, or you may already be knowledgeable. No matter where you or your team is on your AI journey, these studies have been carefully selected to provide value and key insights. Each study will help you understand how others have implemented AI in the real world. Each study will be different, with some being more verbose and others being more brief anecdotes.

Some of the studies will be simple stories. Others will be more detailed and open, with the high-level concepts to set the level of technical depth and perspective of the scenario from the outset. In these more detailed ones, I'll point out some of the nuances that make each example relevant and will also cover lessons learned that you can take away and incorporate into your organization.

One of the biggest recurring themes is how *new* modern training data is. Imagine we are at the dawn of compilers becoming common. Compilers convert high-level code to machine code, so if someone presents a way to manually optimize code,[1] it wouldn't be very useful to a new organization that has access to compilers.

This newness brings many challenges. There are often gaps in knowledge. Case studies can be very hard to come by because of lack of volume, relevance, or confidentiality. This chapter aims to provide new resources, modified and anonymized,

---

1 E.g., branchless conditional

but still based on real stories. One caveat is that since this is a rapidly evolving space, it's fully expected that some specifics may have changed since the case occurred.

For simplicity, I've made up a company name, a fictional "X Company," to reference companies in the case studies with ease. I'll change X to other letters for each study. The letter, e.g., "Y," will be whichever company is in focus in the case study, the real story behind every company is different, and this will help us keep that in mind. All examples in this chapter have been made general purpose to reflect only industry-level knowledge and respect confidentially. Any resemblance to a specific company or context is not intentional, and/or the work is in the public domain.

Industry

All of these in Industry stories and case studies will help reinforce the AI transformation concepts described in earlier chapters. I will provide a wide range of views, from the high-level organizational perspective to the deeply technical team level.

As a preview I'll cover:

- A security startup adopts training data tools
- Quality assurance at a large-scale self-driving project
- Big tech challenges
- Insurance tech startup lessons
- Four brief additional stories

Let's get started.

A Security Startup Adopts Training Data Tools

This story is about a large security startup that found that adopting a training data platform yielded benefits across virtually every team.

Annotators were using an in-house solution that had a bunch of manual forms. They switched to a tool that moved all those forms into a single search box. By being able to search thousands of attributes, across what was previously five plus separate forms, and by having the pre-data loaded when possible, the velocity to annotate improved dramatically. Their internal benchmark showed the improvement to go from minutes per annotation to seconds. The clear general takeaway is that well configured or customized tooling can significantly reduce the number of hours worked.

Another change was replacing the manual file transfer process. Prior, there was a clunky manual file transfer process. This meant that geographically distributed teams often had multiple copies of the data, and the security controls had many holes. By moving to a single standardized training data system, all the security controls were centralized to one place. This also helped reduce data transfer costs, since objects were noted by reference instead of being physically moved. This was sort of like going from manually trucking around gold to modern banking.

Data scientists were perhaps the most to benefit. Before, two different teams treated labels (e.g., with bounding boxes) and attributes as two different things. This meant confusion. By moving to a training data system, they were able to unify their schemas into one. Further, they now query the data directly from the system, and even get notified when sets are ready to be reviewed or progressed.

Finally, by starting to "systemize" everything, they were able to unlock an entirely new product direction that removed much PII and annotation concerns.

In summary, the benefits of their adopting a standard training data system were:

- Security improved, going from A to B to C to D to just one PII-compliant place. Plus, data transfer costs were reduced now that all the data could remain in one place at rest.
- Now data science uses exactly the same schema, instead of three different copies among three different teams, leading to faster model production. Velocity to annotate went from an intractable amount of time to seconds, and "all-new" forms of annotation were now possible.
- An all-new product line was created.

While this may seem "too easy" or overly simplified, this was a real-world thing that happened and was primarily driven by the decision to use a standard off-the-shelf system, and the political capital to drive a core team of change agents to implement and evangelize the change. This is textbook positive AI transformation.

Quality Assurance at a Large-Scale Self-Driving Project

This study is focused on annotation quality assurance (QA) learnings. In this case, the observations are going to focus on the actual QA annotation aspects and will zoom in more specifically on the annotation side. The imaginary "X Self-Driving Company" routinely had to change the schema labels because of confusion. These are some of the QA lessons learned from that process.

The overall theme of this is failure to update the schema. The team kept expanding the instructions and making changes that arguably made the schema worse, instead of holistically expanding the schema to meet the minimum needs of modeling and raw data.

Tricky schemas should be expanded, not shrunk

The model had difficulty deciphering between "Trucks with Trailers" and "Large Vehicles." As a result, the schema was revised, and the team ended up not using the Trailer class, choosing to label them all as Large Vehicles instead. This label included several other vehicles, including RVs, semi-trailers, and box trucks. This reversion to a more generic label helped the annotation effort, but left a huge hole in model performance. My take is that "shrinking" the schema in this context was a suboptimal approach. It would be better to still label something like "Vehicles," and then specify "Size" and "Attachments" as attributes. The "Attachments" section could have an "Unclear" category. That way, the true state of the vehicle can be known. It's worth remembering that we can aggregate labels and attributes during ML training. For example, data science may place "Trucks" and "Large Vehicles" into a single category later as a preprocessing step for the ML training.

Don't justify a clearly bad schema with domain-specific assumptions

At X, small towed objects were labeled as the object towing it. For example, a generator being towed by a large vehicle would get rolled into the "Large Vehicle" label. Their approach was to have the QA team clarify what "Small" and "Large" was. Because their use case was related to semi-trucks, a pickup truck (half ton) was counter-intuitively considered a small vehicle. My take is that this is still a bad schema. The schema should be clear to human annotators and only expert terms should be left for SMEs. This means that it would be better instead to use a label like "Vehicle," and then add attribute types (pickup truck, van, etc.). Having to clarify to everyone that "large" was used in a supposedly use case–specific meaning (large relative to a semi-truck) leaves too much room for error.

Tracking spatial quality and errors per image

To manage quality, track the expected number of errors and normal ranges, as shown in Table 9-1.

Table 9-1. Normal ranges for common pixel and annotation errors (in context of images as the media form)

Expected number of errors	Normal ranges and notes
Of wrong pixels per image	• As far as pixels, this could be as many as 1,000 pixels. • In general, the target was 200 or less, but they would routinely get errors of over 800 pixels. • Setting an accuracy threshold was needed. For example, sometimes QA would try to fix a single pixel on a line, and it was just not an effective use of time.

Expected number of errors	Normal ranges and notes
Of wrong annotations per image	• They would generally expect less than 0.02 errors per image. • 0.1 errors per image was a cause for major alarm. • Practically, this would mean that for every 50 images, which may be over 500 instances, there should only be 1 error. Note that they would set a threshold for this above a certain pixel value. So, for example, "corrections" of less than 50 pixels or 50% change (whichever is greater) may still be required, but not counted as "errors" for the sake of this metric.

The level of quality from the provider ranged from 47 to 98% depending on the label. In general, with more severe effort from the customer, the customer could self-audit to 98%+. In general, it appeared that getting above 98% was almost certainly in the realm of different opinions or guesswork, and was more than what could actually be determined from the image.

In general, there was a relationship between volume and errors. Usually, the higher the volume, the lower the relative error. In theory, this could be because annotators were getting more familiar with the data and self-correcting.

I have two main takeaways:

- Surprisingly, at least to me, it is possible to track quality using this method.
- 98% or above is as "perfect" as you can get for spatial-related work.

Schemas for lane lines need to represent the real world. Company X had recurring trouble with lane lines. While the images can't be reproduced, visually I would say that part of the issue is that lane lines were simply very ill-defined. For example, lane lines appeared with a smidge of black lines (not white), there were painted-over lines, and lines where the "ghost" of the paint was there, but there was no actual paint. My take is that they did not align the schema to the raw data well in this case. It's not that hard, looking at it, to iterate out some of these common cases, but instead, they would blame the annotation team (instead of fixing the schema).

Regression and focused effort do not always solve specific problems

Continuing the example of the lane lines, they found that focused attention did not automatically solve the problem. For example, during weekly reviews, they would highlight a problem area, and that problem may actually get worse the next week.

To highlight errors (e.g., of the lane lines) they would use a comparison format, where the image was duplicated with the fix shown on the second image. This worked fairly well, but the degree of error was hard to understand because the zoom/crop rate was variable. This meant that sometimes an error looked small when it wasn't, or vice versa. Including some kind of mini-map or zoom percent may help alleviate this.

Overfocus on complex instructions instead of fixing the schema

A recurring theme in this project was a failure to fix the schema, as they instead attempted to solve it by ever more specific, complex, and frankly confusing instructions, as shown in Table 9-2.

Table 9-2. Contrast of labeling instructions changing over time with author's commentary

Initial instruction	Problem	Second instruction	Author's commentary
"Don't label faded lane lines."	Model struggled with images with faded lane lines.	"Label all lane lines no matter what."	Label all lines AND put an attribute of "faded" on faded lines. If needed, de-emphasize overall volume to ensure critical quality threshold is met.
"Do label all side guards even if you can't see them."	Model overpredicted side guards.	"Only label what you can see."	Only labeling what you can see is usually the best default starting point.
Label "vegetation" differently from grass and weeds, etc.	Annotators easily confused "vegetation" with forms of vegetation such as grass and weeds.	Not addressed.	Always be clear about what is an aggregate and what is a specific thing. Rather, have "vegetation" as a top-level label and then [grass, weeds] as attributes.

They also had trouble with the drivable road surface being different from the background (e.g., a road that's visible, but not accessible from the current place). Again, this is not that hard to think about conceptually; even simplistically, this can be two different things, but instead of changing the schema, they just kept growing the instruction set (it literally became a Wiki with hundreds of pages, which is just crazy!). Yes, good instructions are important, but they should not replace having a good schema. Always try to correct problems in the schema itself before changing instructions. An analogy is like "correct the form on the website instead of adding more and more instructions about how to use the form."

Trade-offs of attempting to achieve "perfection" in nuanced domain-specific cases

One of the themes was tension between wanting to achieve a high degree of quality, but also not wanting too much time spent on relatively small corrections. At one point, a limit was set that if greater than 4× zoom was used, they were to not make the correction.

They did not do a good job separating purely observable information from what we as humans were conceptually adding. For example, a human might guess that a mailbox (or "x" thing) is beside the road, so it's labeled as such, but the observable image (without the assumption of the context) is just a dark patch that doesn't show anything.

Another tension point was that sometimes the "what can you tell from visual only" was expected to include a near-complete driving experience, with knowledge of the laws of the local region. For example, to determine if it's a road or a shoulder a person must sometimes know the meaning of line colors and types, and the expected width of the road in cases where the road surface is larger than reasonable and part of it is considered to be the shoulder.

The nuanced cases in particular are very difficult to determine a fixed solution for because they come across as arbitrary in many cases. In general, it seemed very difficult to achieve a high accuracy, at a reasonable QA cost, and especially with the borderline and nuanced cases. Let's take a closer look at some examples of these nuanced cases.

Understanding nuanced cases

In the cases where there was occlusion, it was deemed OK to have a continuous label (overriding the visual evidence requirement). However, the trick was that every extent had to include known information. For example, a lane that starts on the left and is visible, then is blocked by a truck, then is visible again, was OK. But a lane that is only visible at the start or end was considered not OK.

Distinctions between easily confusable things caused a lot of issues; here are two examples:

- "Paved" versus "unpaved." This is more ambiguous than it sounds. For example, is stonework considered "paving"? Is a lane on the right of a solid white line road surface or shoulder? In California, for example, the answer to that is ambiguous in many cases (for example, dual turning lanes).
- "Terrain" versus "shoulder" was overloaded (multiple meanings) and ill-defined. For example, many roads have shoulders that are essentially also terrain, so the term is overloaded. Guardrail sections that intersect with other labels face a similar problem. For example, the few pixels of ground a guardrail is stuck into would probably be better labeled "guardrail" or "guardrail base," since you can't drive into that space without hitting the guardrail!

The value of tracking "pixel-wise" error seems to be questionable because it's highly correlated with the class (e.g., terrain has 100× more pixels than a railing). It seems that method is only useful if scoped to the same class.

Stacking ranking errors by class did appear to highlight the most common errors, but the practical value of this appeared to be somewhat limited.

Learning from mistakes

Here, I will recap some of these mistakes and offer my take on better approaches.

Define occlusion well. Occlusion was poorly defined in general. For example, a transparent fence was to be called the underlying class. A better choice would be to have two labels. A commonly confused thing is that people think that you can only have one label for a given "slot" of space. That's not true. You can have overlapping labels. You just need to specify the occlusion or z-order (which in turn can be predicted by the ML system). So in this case, there should be a label "fence," which could also have an attribute "foreground" or "most visible." Then there is a second annotation "road," which contains the attribute "background."

Expand schemas. They used overly broad class definitions. For example, is a truck-mounted RV a "large vehicle"? Yes, by most measures, but not for their domain-specific (semi-trailer size) area.

A better approach is to have a much more diverse set of annotation classes (e.g., using labels and attributes) that allow the annotator to be as accurate as reasonably possible, and then have a set of relationships back to whatever label you want to train on. So the annotators should still just say "RV," and then the system can map that to "large vehicle" if that's how you want to train it. Trying to fix overly broad classes with instructions is jamming a square peg into a round role.

Remember the null case. A general theme was that "things outside the immediate view" are hard to label. Is that blur of gray and green on the extreme edge of the camera vegetation or buildings? I don't know. In general, they seemed to be missing "unknown" or "null" classes.

Another example was the difference between green hills in the background and the "vegetation class." The green hills *are* vegetation, but the request was to label the hills as background because they (using knowledge outside of what is visually obvious) are farther away.

Missing assumptions for language barriers. Keep in mind that annotators may not be native speakers of English, so a distinction like "dashed" versus "dotted" lines may be lost in translation or lack of background context. Also, for most of the images I saw, at the resolution being used, "dashed" and "dotted" was ambiguous, so a person could argue either way, meaning this was probably a bad class name. While local or in-house subject matter experts may not have an English language barrier, the domain knowledge and domain-specific terms form a language barrier in and of themselves and so this problem remains.

Don't overfocus on spatial information. Sometimes, too much attention was paid to spatial location in some class types. For example, for an advertising sign (not a legal road sign), they would expect the support posts to be labeled differently from the "sign" itself. Whether this was needed for the ML model, I don't know, but it likely took ten times as much time to annotate every sign, and made QA harder. My two cents is that would also make it much harder for a model to predict it, since it would take what otherwise could be a bounding box and turn it unnecessarily into a segmentation problem.

Big-Tech Challenges

This study is zooming out and looking at the broad organizational effect around training data. "Y Company" (a fictional company) is a leading producer of consumer electronics and has a massive AI organization.

There are many talented people at Y Company, but their overall production of training data is lagging behind competitors. Despite being the best in many other areas, their AI product is not competitive with other products in this space. They have decided to make some changes to fix their organization and, while the results of those changes are yet to be seen, I think the reasons for those changes and desire to make them highlights some important training data lessons.

To set the stage, Y Company follows the textbook examples of the multiple small teams approach, where the overall problem is divided into a variety of sub-problems and each team works in a specific area. For example, there are data engineering (infrastructure) teams, annotation team(s), data science teams(s), etc.

Let's take a closer look at their structure and some of the challenges they encountered.

Two annotation software teams

In part because of the breadth of the space, Y Company has two different annotation software teams.

Each team is responsible for the entire end-to-end annotation process, including software tools. One team covers some interface types, such as images, whereas the other team covers different types, such as audio. The responsibilities look something like Table 9-3.

Table 9-3. Two annotation software teams doing similar tasks

Team one	Team two	Overlap?
Audio interface	Image interface	No
Import/export	Import/export	Yes
Storage abstractions	Storage abstractions	Yes
Human workflow	Human workflow	Yes
Automations	Automations	Yes
Third-party integrations	Third-party integrations	Yes
Scheduling, general operations	Scheduling, general operations	Yes
User admin, overall management	User admin, overall management	Yes
Hardware infrastructure	Hardware infrastructure	Yes
Much more...	Much more...	Yes

Can you spot the difference?

They realized that the teams were doing very similar work with only the surface level UI being different. And clearly, having multiple software teams doing essentially the same thing is suboptimal (think 95%+ overlap). After much effort, they started a multi-year project to merge the systems.

But wait, you may say—surely there are valid reasons for this? Integration with audio must not be the same as images? That argument, in this overall system design context, confuses the fruit with the fruit cart.

The overall ingestion system, and high-level integration concepts around shared principles like labels, attributes, storage adapters, etc., is the cart. The specific data type that will naturally have a different format is like the type of fruit. You set up a route, and a driver, and get the cart once, and then can transport different kinds of fruit.

Shifting to the context of using off-the-shelf systems, for a very large setup, the hardware configuration of the audio-focused instance could be different from the image-focused instances. But that's a configuration detail, not each team making totally novel independent design decisions on hardware.

For human tasks, yes, there are some practical differences between a long-form media type like video or 3D, and certain more simple types of images. But this comes back to UI customization, and relatively small changes. The core principles around task management, user management, etc. remain the same.

Again, this may seem obvious in retrospect, but if the project starts as a "get it done quick" order from data science, and there are apparently narrowly defined data types, labels, volume of users, etc. it's relatively easy to miss.

Confusing the media types

To put it plainly, having two similar platforms is costly.

Since this seems so obvious in retrospect, I will explain how it appears to have come to pass at Y Company. Chiefly, this is a confusion between the UI concerns and the platform needs.

When these projects start, the annotation interface (the UI) is rarely thought of as a *platform*. People get fixated on things like the end-user-facing interface (images, audio, etc.) and forget about the vast amount of behind-the-scenes work to arrive at that.

The clear mitigation step here is to think of a training data *platform* as the first choice, and the specific interfaces that it uses as a second choice. The trend currently is for platforms to support all popular interfaces anyway, so if you are buying an off-the-shelf platform, this may already be addressed.

Why does the platform concept come into play? Because, as I showed, all media types have similar underlying challenges. Be it images, video, or something else, it all needs ingestion, storage, human workflow, the annotation interface, the automations, the integrations, the connections to training, and more. The media-type-specific interface is a sub-concern of the annotation interface area.

Non-queryable

Y Company has a dedicated team for different layers of the process.

Unfortunately, in this case, the different layers were defined in terms of data science steps without regard to the centrality of training data.

Data science is in a tough spot, because much of the metadata around objects is not accessible to them upon query. Essentially, this means they had to query each object, get it, and inspect it, in order to then build their dataset, as shown in Figure 9-1.

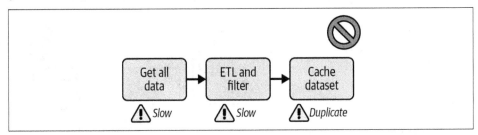

Figure 9-1. The slow and error-prone ETL process at Y Company

There are several practical implications:

- Most of the security controls designed by the import team were routinely defeated. Data that was supposed to be deleted in a few years would last for much longer in the data science team under the guise of datasets.

- At this scale (think 10s–100s of millions of records) there was no hope of any speedy updates. This meant that they may need to pull and store a million records, in order to get the few thousand they wanted.

- This put huge burst load problems on the storage team, since they were expected to support absolutely massive on-demand access (think <10k QPS and >500 modes).

While some level of "loading extract and transform" (or the various orders of those concepts) is standard, in Y Company's case, it becomes excessive and without clear value added. The better approach is for the prediction and raw media data to go more directly to a known format and be pulled and queried from a shared known format and logical (or even physical) place.

Different teams for annotations and raw media

For data science to actually get the raw data, they had to talk with two different teams. First, they would get the annotations from the annotation team, and then usually they needed to separately get the raw media from the storage team. This led to inevitable problems like having to merge records and finger pointing. They would have the ID for the annotation, but not the raw data, or vice versa. Deletions were a huge problem. A record may be deleted in one system, but exist in the other.

This means that the process shown in Figure 9-1 was sometimes further duplicated, or additional processes must be created to broker or merge between them, as shown in Figure 9-2. This also caused issues in reverse, e.g., that the training data team needed to correlate and load predictions and then merge them with raw media.

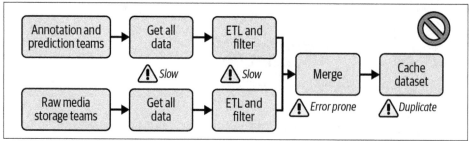

Figure 9-2. Multiple teams exacerbating ETL process further

Part of why this was such a problem is that there was no clear upstream. No clear flow of data. No clear producer and consumer.

This resulted in major issues, both of which remain largely unsolved at Y Company:

- Data science is forced to "greedily" extract data so as to ensure they can actually get it. Basically, they can't "trust" the upstream system, so they are routinely caching the data. And in fairness to the raw storage team, it wasn't that they were doing anything wrong, it's just that they didn't control the annotation. This was an organization mistake—no single team could correct it on their own.

- Annotation (teams) have to use the storage as an unnecessary proxy to access data. They often end up caching the data too.

Moving toward a system of record

It is very clear there is a desire to move toward a more unified approach, a single system of record.

For example, training data teams can access the storage layer directly instead of the unnecessary proxy. At the time of the case study, Y Company was evaluating this training data database as this system of record, with key concepts shown in Figure 9-3.

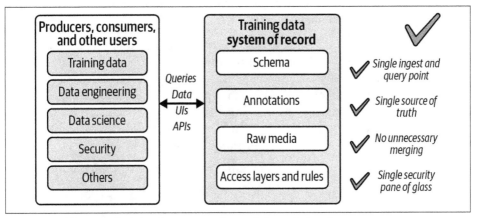

Figure 9-3. Users and goals of a system of record for training data

There is a shared state of annotations and raw media. This means that when data is ingested (for example, a production-level prediction), that representation is framed in the same way that the annotation team will frame it. A record that's stored as raw storage will be accessed directly from annotation.

This also changes the reporting relationship. There can now conceptually be a flow of training data producing and data science consuming. In practice, this does not mean there is a massive data pipe going from training data to data science. Rather, think of it as a notification. Data science can track the activity of the training data production, but when they pull data, it's from the same source as that annotation wrote to.

To understand why training data becomes the center of gravity for these systems, consider these specifications:

- Data science must export the data, on an ongoing basis, in order to use it.
- The formats, privacy rules, data, and organization change routinely, sometimes hourly or daily.
- Data science consumes training data and produces predictions consumed by training data.

Missing the big picture

Despite Y Company having many oddly specific things as top-level "named concepts," training data was not previously one of them. This confusion was at the heart of the issue; there was not an overall concept of training data, with each team doing their own data processes, and no "big picture." As training data became a named concept, the pattern of having a system of record for it became more obvious. This changed what was previously a total mess into the earlier shown Figure 9-3.

This also helped create clear streams, such as from input to training data producers to data science consumers. A key point of consideration for careful readers of this book is that while conceptually, training data produces and data science consumes, at a large organization, there are many other processes and loops to consider. For example, data engineering may have production predictions that are fed back into training data. A security team may desire a "single pane of glass" for the data. All of this is relatively straightforward with a single system of record.

Solution

The solution consisted of two stages:

1. See the big picture, see the training data mindset, and observe the overlaps between multiple teams.
2. Move toward a system of record to align multiple teams.

While the new direction was still a work in progress, the expected results of this change at the company were impactful:

- There is now a single source of truth (including ingest and query points): The data science team accesses the latest version of the data. The need of data science to query the data, form named datasets, create immutable versions, etc. are all given first-class consideration. This reduces the otherwise painful overhead of trying to fight to get the data, converting it into a smooth process. An analogy here is like querying a database versus having to get an FTP transfer and then reconstruct it.

- There is no longer unnecessary merging: The data is captured at ingest in such a way so as to support annotation from day one. This means that the need to annotate is again a first-class consideration. This eliminates the need for the annotation team to have to create their own methods here. To be clear, there can still be a team of people working on the literal interface, but they are using access methods defined by the training data system, not creating their own secondary ingestion process into their system.

This also creates much more linear reporting relationships. If a new production of data is needed, it can be clearly defined and reasoned about instead of having to nebulously coordinate between three or more somewhat opposing teams.

Let's address loops

This takes nothing away from pre-labeling, or other interactions, iterative improvement, etc. If anything, by creating a closer relationship between training data and data science, it enhances the ability to improve these automations. If it helps, you can imagine there being a loop between producers and consumers, and/or between the input and output. The preceding diagram is "unrolled" so as to clearly indicate the key relationships.

Human in the loop

One of the other confusions here is that there is sometimes the need for data to be graded by humans in near real time, or in other ways to use that grading loop. In the cases where that is truly necessary, presumably data science or data engineering is aware of this, and must still pass through them in some capacity. For example, just because a human annotator supervises a data point, that does not mean that exact output is what is presented to an end user. Normally some other process must run to transform that output. For example, the annotator may correct a label, but the end product doesn't present labels, it presents aggregated statistics or alerts, etc. There may be a literal direct technical integration from annotation, but the preceding diagram is more in the spirit of the human organization.

The net effect here is that there is a lot of data duplication. My take is that part of the problem was not having training data as a named concept. Instead, Y Company was too narrowly focused on the business stakeholder goal without thinking about the wider context in the company.

The case for aligning teams around training data

We covered a lot of ground in this case study. Let's review our findings. A tech company had dedicated teams for different layers of the process, like an "ingest" team. This meant that often, different teams within the company would treat it as an unreliable source, hoarding the data for their own use. This meant that most of the security controls designed by the import team were routinely defeated. Data that was supposed to be deleted in a few years would last for much longer in the data science team under the guise of datasets. There is now a desire to start moving toward a more unified approach, where the same system manages both the dataset organization and the import identification.

A rough analogy of this is the cliche about microservices where three different teams all have a 300MS speed target. But then when the customer actually uses it, the services have to talk to each other multiple times, meaning that the actual input/output of the system becomes something like 5–10 seconds or more.[2] A similar thing happens with training data, with each team giving service targets that have little relation to the goals of the other team. For example, data science is going to care about dataset level access and queries. But if the data is provided by the import team in a way that is not queryable, then data science must get *all* the data first. Keep in mind that at this scale, we are talking about hundreds of nodes, so this is a major operation.

There are a couple of lessons to take from this. One is that treating each layer of the stack too independently is worse than treating it as a single end-to-end process.

The main lesson here is to treat training data as a first-class named concept, and then lean on normal concepts like having a system of record for training data. The company is doing this, and already merged both of the teams, and is continuing to drive toward a system of record used by multiple teams for training data–focused concepts.

Insurance Tech Startup Lessons

This story is looking at the effect of training data on the life of a small startup.

A notable insurance tech startup used photos of accidents to improve the automatic claims process. I discussed the failure with the chief engineer of the team, and the primary conclusion from that discussion follows.

---

2 E.g., IN->A->B->A->C->B->OUT

Will the production data match the training data?

First and foremost, the insurance adjusters were not ready for AI adoption. This had the immediate technical impact that they would take "bad photos." While some of the training was high resolution, and in some cases video walk-around, the photos they had to use in production were of a much lower quality.

Unfortunately, there was not enough political credibility to invest further efforts after the initial production attempts failed. This may have been avoidable if a deeper understanding of the expected production photos was gained before stating the system was ready for production.

To use a more classic system analogy, the availability of production data needs to be considered at the earliest architectural stages of the system. If you leave it to the user acceptance phase, then there will be big trouble—it's not a few UI pages you can correct at the last minute, it's the core of the system.

The main lesson is to make sure that the production data will match the training data.

Too late to bring in training data software

The chief engineer had been pushing for use of commercial training data software for years. His efforts fell on deaf ears at the C-suite level, because at the time, the focus was on their novel augmented data method.

Around two years into the project, after it became clear that the augmented method was not working, the CEO revisited this issue with new guidance to look into training data software. At this point, it was too late, but the lesson was clear. Bring training data software early to a project. Many startups have tried to "beat" the system by skipping annotations and to my knowledge, not one has succeeded yet.

Stories

The following are four brief stories of training data in the wild. While not full case studies, they provide interesting and useful anecdotes to help frame your work. The stories are:

- "Static Schema Prevented Innovation at Self-Driving Firm"
- "Startup Didn't Change Schema and Wasted Effort"
- "Accident Prevention Startup Missed Data-Centric Approach"
- "Sports Startup Successfully Used Pre-Labeling"

"Static Schema Prevented Innovation at Self Driving Firm"

This story looks at a large self-driving car company and the effect of a lack of awareness of training data. A prominent self-driving company was using static labels. Something like "Car" or "Bush" was a static concept to the system, meaning that any time they wanted to add a new label, it was a huge effort.

In comparison, another firm had taken a more data-centric approach to these labels, allowing for the introduction of concepts like "Cut-off Detection" (when a driver gets "cut off" by another driver in traffic) more smoothly. While this is naturally a complex problem, the carmaker with the static named concepts has missed many deadlines, while the one with the more flexible approach to labels has cars on the road today. Some lessons from this: set up the mental models, tools, and processes so that you can be flexible with your training data. Explore and change the label schema as the situation warrants. A flexible and growing schema is a must-have for any modern training data project.

To understand this better, a key blocker was that the way the org structure worked, the training data team focused just on the infrastructure, and was at the mercy of other teams for the definition of their product (including concrete predefined schema names like "Car").

At the time, the demand from these other teams was around how to serve the data efficiently to their training system—not working with them on a flexible and dynamic schema. This lack of cross-team awareness and discussion of training data led to a lack of flexibility in model predictions and a more general stagnation of the tech.

This was a classic example of being too busy to try something new, as shown in Figure 9-4. It's a good reminder to not be too attached to current ideas and solutions even if they appear to be working. Instead, it's better to have a training data first mindset.

Figure 9-4. Cartoon showing over-focus on day-to-day work over new innovations

"Startup Didn't Change Schema and Wasted Effort"

This story is about a company that had kept the same fairly simple (mostly top-level labels) schema for about two years. Every day, a team of 10+ annotators dutifully annotated new examples from scratch (not even pre-labeling most of the time).

They had no real benchmark to realize that this was really bad! If the model can't detect something after a few months of annotating the examples, one million more examples is probably not going to help.

Taking a closer look at their data, it became clear that they were annotating an object X that looked completely different in different images. That meant that they were really confusing the model more than they were helping it. We improved their performance by adding in tens of new attribute groups that helped them zero in on model performance.

"Accident Prevention Startup Missed Data-Centric Approach"

This story is looking at the effects of a mid-sized startup struggling to grow from a company that makes dashcams to prevent accidents. The company was "high flying," but stagnated.

The company did not adopt a data-centric approach. The company had very few annotators and a relatively large engineering team. This meant that a lot of effort went into classic "engineering" approaches to solve problems that would have been easily solved with some annotations. Because they were paying about 20x+ more for an engineer over an annotator, this was especially problematic. This mainly appeared to be a lack of knowledge of a training data mindset.

Regarding tooling, the company acquired another small startup that had an in-house annotation tool as a side project. As you can imagine, this was not a robust tool, but they felt committed to it and wouldn't change for more modern tooling. This gap seemed to be from a lack of understanding of the big picture of training data and the cross-functional importance of a system of record for training data.

On use cases, they over-focused on the type of annotation method (e.g., image segmentation) instead of use cases. They had ample opportunities to expand existing use cases by adding deeper attributes to existing bounding boxes, but missed it by getting blinders on regarding the spatial detection concerns.

Overall, there was little internal excitement about the company's current roadmap, and lack of product innovation was part of it. Given how heavily their product leaned into image use cases (from the road ahead, to driver monitoring, etc.), it was a big missed opportunity to lean into a training data approach.

There are few key takeaways:

- If your engineering team to annotator ratio is positive, that's bad, and an indication that you probably aren't using a data-centric approach.
- In-housing training data tooling is usually suboptimal; you should start with open source tools and build from there (and ideally contribute back to them).
- Use the Use Cases rubric in Table 7-5 to understand and scope better use cases.

"Sports Startup Successfully Used Pre-Labeling"

A sports analytics company has used pre-labeled data on videos to great effect. They managed to essentially reduce the box drawing time to approach zero, freeing up annotation time to work on attributes that would directly add value to their application. This effort was measured on the order of years.

I include this brief story to note that it is possible to have success with automations. Also, most of the effort over time was in annotations and sequence/event predictions, not on spatial locations.

An Academic Approach to Training Data

All of the previous studies and stories were from industry. I'd like to wrap up this chapter by talking about an academic challenge called the Kaggle TSA Competition.

Kaggle TSA Competition

In 2017, Kaggle, a data science competition platform, hosted its largest ever global competition: 1.5 million dollars in prize money, 518 teams of some of the top minds in the space. I participated in this competition too, so I can also speak from the participant's perspective. To this day, it remains one of the competitions that has attracted the most attention ever.

Officially, the theme was "Improve the accuracy of the Department of Homeland Security's threat recognition algorithms." The technical concept was to detect threats at airport security based on data from a 3D millimeter wave scanner, as shown in Figure 9-5. The data was provided in pre-processed slices of 64 images per scan, meaning that in general, most people treated it as plain RGB images. The principal goal was, given a novel set of scans, to detect with a very high degree of accuracy what threat was present and where. The prediction goal included localizing to 1 of 17 location zones. These zones were abstract, however, and did not reflect any kind of existing spatial training data.

Next time you are at the airport, if you look at the display beside these scanners, you will likely see it's mostly empty, with a general "OK," or a box on a generic model of a person. In this case, an "OK" means a super-high confidence that there is nothing present. A box on a specific location means a "probably nothing, but not quite enough to pass without additional check." The vast majority of people are allowed to pass. And the goal is to minimize manual intervention to improve process efficiency as well as the passenger experience. Achieving this goal means highly reliable algorithms with a very low false positive or false negative rate are needed.

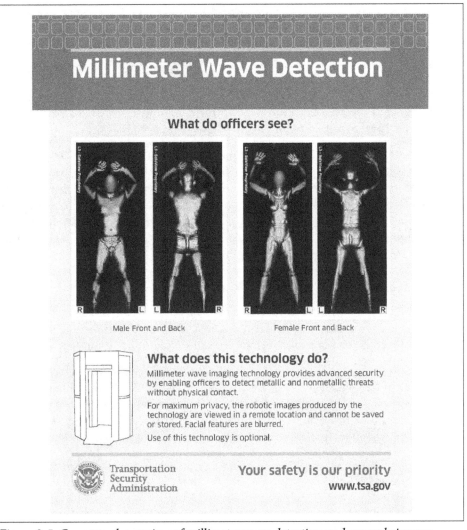

Figure 9-5. Conceptual overview of millimeter wave detection and example images from public domain (The image is in the public domain (https://oreil.ly/B91iH) in the United States.)

Keying in on training data

Training data was what mattered, and every top approach focused around it. The winning approach centered around data augmentation. Many of the top-placing approaches used new human-made annotations that captured spatial data. To capture an idea of how intense the competition was, one competitor missed out on a $100,000 place, because out of thousands of classifications, a single misclassification was included in their results.

That being said, the majority of approaches within the top 10% were actually fairly similar in performance, with the winners mostly doing competition-focused work like multimodal ensembles. Further, in that context, it was possible to get in a top rank simply by creating training data annotations for it. One approach was to manually annotate the data using bounding boxes (over and above the provided data). This newly annotated spatial data allowed me to create an extremely simple model setup. One model predicted if a threat was present, and the other predicted where on the body the threat was.

To try to frame the context of this, while it's common knowledge that actually winning a Kaggle competition often falls down to hyper-specific things, usually these things are to do with data science itself (e.g., model parameters, etc.), whereas in this case, it was much less about the models. All of the winning write-ups I have read have all stated most casually what actual model architecture they used. This is foundationally different from most other cases, where the model choice, feature selection tuning, etc. mattered most.

How focusing on training data reveals commercial efficiencies

One of the common criticisms of these types of competitions is that the winning approaches are impractical for actual use. Often, they involve orders of magnitude more computation than is reasonable, or over-fit dramatically to the specifics of the way the competition frames the problem.

For me, one of the biggest wake-up moments came back in 2017 when I realized that by using an "off-the-shelf" object detector, with the only real difference being the training data, I could in a matter of days beat or at least match virtually every other method. This was about as close to a legal cheat code as you can get.

To circle this back to the commercial efficiency, it meant that instead of needing to treat each dataset as this unbounded scientific research initiative, it could be reduced to the calculation of how much annotation effort was needed, plus some reasonable estimate amount to train the model. While this glosses over a lot, fundamentally it remains true today and is at the heart of the swarm of commercial interest in this technology.

Learning lessons and mistakes

Up to a certain point, adding new training data was like magic to improve performance. After a while, it reached a point of diminishing returns. These are some of the lessons I've learned, and reflect what I would do differently today for a production system.

I only annotated the high-level category of "threat," which meant it was hard to know how to improve the model for specific types of threats. Adding more training data to cases that are already well-covered does little. For example, I recall there were certain categories that were much harder than others. I manually shifted my efforts to focus on samples with the worst performance. However, this was all just ad hoc. Knowing what I know now, I would create a schema that addressed those cases directly. That way, I could statistically know how well each case was being covered, versus going with a "gut feel." In other words, a production version of this would need a much more detailed schema.

Bad training data is incredibly and disproportionately damaging. After a while, I realized that sometimes an image was being included without proper labels, or with inconsistent or wrongly placed labels. I fixed these errors after I realized what a negative effect they were having. A good rule of thumb is that a bad example is only offset by three or more correct examples.

There are many articles and books on how to improve model performance. I specifically want to highlight again that all of these points center around training data. It doesn't matter what type of model training process you are using; all of the preceding lessons will still apply in one or another.

Summary

There are a few key concepts and recommendations I hope you take from these case studies:

- Static schemas limit innovation. Flexible, growing schemas are important for training data.
- Raw production data should be similar to raw data in training data datasets. Consider this early in system design.
- Thinking they can "engineer their way out" instead of using a training data-centric or data-centric approach is a common mistake for startups.
- For large organizations, aligning teams around a training data system of record is better than slow, duplicated, and error-prone "free-for-all" approaches.
- Bad examples can disproportionately damage models. Track issues within specific schema attributes to learn from mistakes and more easily continuously improve training data.
- Thoughtful design of schemas, workflows, and quality assurance is important for overall success.

The recurring theme is that real-world training data is nuanced and complex. Hopefully, these experiences have given you some insights to apply to your own contexts.

Index

dataset connection, 150-152
processes, 150
dataset connection, 150-152
streaming, 149
DataMaps, 114
datasets, 147
definition, 16-18
versus human supervision, 173-176
manual creation, 189-191
pipelines, 150-152
processes, 150-152
transfer learning and, 178-180
debugging, ML modeling and, 146-147
decision making, scope, 53
dedicated tooling, 63-65
definitions, attributes and, 70
development system, 43
sequentially dependent discoveries, 43
Diffgram, 33
direct supervision, human supervision, 27
Director of AI Data Responsibilities, 218
disambiguating storage, 119
discovery, 175
versus automation, 174
Docker, 48
domain-specific automation, 270-271
domain-specific query languages, 36
dynamic organization, 149

E

efficiency, 12-13
ellipses, 84
embedded supervision, 39
employees, cost center, 38-39
end users, 17
input, cost reduction and, 42
engineering, data representations and, 76
errors in data, 140
common causes, 141
ethics, supervised data, 181
event-focused analytics, 113
events, interactive automations, 256
exploring data, 144-146
exporting
data access and, 119
file-based exports, 119
external references
connections, 112

F

failures, 23-24
deployed system example, 24-25
file-based exports, 119
files
compound, 114
user-defined types, 114
filters, 149
folders, 149
form type, attributes, 70
full image tags, 83

G

game of telephone analogy, 101-103
naive approach and, 105
GenAI (generative AI), 25-28
General Language Understanding Evaluation
(GLUE), 197
generative AI (see GenAI)
generative pre-trained transformers (GPTs), 25
generic attributes, 71
geolocation, 110
geometry, keypoint, 86
geometry-based labeling, 270
geospatial analysis
BLOB (Binary Large Object) data, 114
GeoTiff, 114
GLUE (General Language Understanding Evaluation), 197
Gold Standard training, 136
GPTs (generative pre-trained transformers), 25
GrabCut, 257
guides, 88-89

H

HCI (human–computer interaction), 39
HCS (human computer supervision), 39
heuristics-based labeling, 271
historical data
theories, 169-170
human alignment, 27
(see also human supervision)
human computer supervision (HCS), 39
human supervision
constitutional set of instructions, 27
direct supervision, 27
GenAI and, 26
indirect supervision, 27

push automation
   cost reduction, 42
Python, 170

Q

QA (quality assurance), 239
qualitative evaluation, 177
quality assurance
   annotators
      as partners, 139-141
      trust, 139
quality of data, 7-9
   annotations, 8
quantitative evaluation, 177
query languages, domain-specific, 36
querying versus streaming, 118
question attributes, 70

R

raster masks, 85
raster methods
   semantic segmentation, 90
raw data, 5-6, 96-97, 195
   data relevancy and, 177
   media types, 22
   spatial type and, 194
   storage, 109
      geolocation, 110
      pass by reference, 110-111
      pass by value, 110-111
      storage class, 110
      vendor support, 110
RBAC (role-based access control), 121
relationships, 87
   when, 87
remote storage, 116
research sets versus applied sets, 197-198
ROI (return on investment), 48, 230
role-based access control (RBAC), 121
root-level access
   security, 59

S

SaaS (software-as-a-service)
   versus installed training data, 42
SAM (Segment Anything Model), 83
sample creation
   binary classification, 188

classification upgrade, 192
geometric representation, 187
manual set creation, 189-191
strawberry picking system, 186-187
scale, 44
   defining, 44-46
   large scale, 46-48
schema, 4-5, 67, 133
   attributes, 67
   automation, 241
   complexity, attributes and, 72
   depth, 193
   design failure, 5
   joint responsibilities, 82
   labels, 67
   persistence, 134
   spatial representations, 67
   system usefulness theory, 166-167
scope, 51
   attributes, 70
   cautions, 54
   decision-making process, 53
   platforms, 52-53
   point solutions, 54-55
   suites, 52-53
scoping automation, 247
security
   access control, 121
   architecture, 57
   attack surface, 57
   authorization, 121
   configuration, 57
   data science access, 59
   identity, 121
   installed solutions benefits, 58
   PII (personally identifiable information),
      124
   root-level access, 59
   URLs, signed, 122
      cloud connections, 123
   user access, 58
Segment Anything Model (SAM), 83
segmentation
   SAM (Segment Anything Model), 83
   semantic, 90-92
semantic segmentation, 90-92
separation of end concerns, 40
sequences, 87
   when, 87

tracking objects through time, 158

About the Author

Anthony Sarkis is the lead engineer on Diffgram Training Data Management software and founder of Diffgram Inc. Prior to that he was a software engineer at Skidmore, Owings & Merrill and cofounded DriveCarma.ca.

Colophon

The animals on the cover of *Training Data for Machine Learning* are black-tailed prairie dogs (*Cynomys ludovicianus*). While they are actually a type of ground squirrel, they received the name prairie dog because of the habitats they live in and because the sound of their warning calls are similar to a dog's bark.

Black-tailed prairie dogs are small rodents that weigh between 2 and 3 pounds and grow between 14 and 17 inches long. They have mostly tan fur that is lighter on their bellies and their namesake black tail tip. They have short, round ears, and eyes that are relatively large in comparison to the size of their bodies. Their feet have long claws, which are ideal for digging burrows into the ground.

True to their name, black-tailed prairie dogs live in a variety of grasslands and prairie in the Great Plains of North America. Their habitat usually consists of flat, dry, sparsely vegetated land, such as short grass prairie, mixed-grass prairie, sagebrush, and desert grasslands. Their expansive range is east of the Rocky Mountains in the United States and Canada to the border of Mexico.

Black-tailed prairie dogs may not be considered endangered, but they are a keystone species. They impact the diversity of vegetation, vertebrates, and invertebrates because of their foraging habits and presence as potential prey. It has been shown that grasslands inhabited by them have a higher degree of biodiversity than grasslands not inhabited by them. Prior to a large amount of habitat destruction, they used to be the most abundant species of prairie dog in North America. Many of the animals on O'Reilly covers are endangered; all of them are important to the world.

The cover illustration is by Karen Montgomery, based on a black and white engraving from *The Natural History of Mammals*. The cover fonts are Gilroy Semibold and Guardian Sans. The text font is Adobe Minion Pro; the heading font is Adobe Myriad Condensed; and the code font is Dalton Maag's Ubuntu Mono.

Printed in the USA
CPSIA information can be obtained
at www.ICGtesting.com
JSHW052005131123
51999JS00003B/6

9 781492 094524